T0185670

Big Data 2.0 Processing Systems

Sherif Sakr

Big Data 2.0 Processing Systems

A Systems Overview

Second edition

 Springer

Sherif Sakr
Institute of Computer Science
University of Tartu
Tartu, Estonia

ISBN 978-3-030-44189-0   ISBN 978-3-030-44187-6   (eBook)
https://doi.org/10.1007/978-3-030-44187-6

The first edition of this work was published in the series: SpringerBriefs in Computer Science, with the subtitle: A Survey

This Springer imprint is published by the registered company Springer Nature Switzerland AG.
The registered company address is: Gewerbestrasse 11, 6330 Cham, Switzerland

*To my wife, **Radwa**,*
  *my daughter, **Jana**,*
  *and my son, **Shehab**,*
  for their love, encouragement, and
support.

Sherif Sakr[1]

To Sherif, may your soul rest in peace.
  Your last words were "Go on." So we do
and will never forget you.

Radwa, Jana, and Shehab

---

[1] Sherif passed away on 25th of March, 2020. As this is one of his last books, his family wrote a dedication to his soul.

Foreword to the First Edition

Big Data has become a core topic in different industries and research disciplines as well as for society as a whole. This is because the ability to generate, collect, distribute, process and analyze unprecedented amounts of diverse data has almost universal utility and helps to fundamentally change the way industries operate, how research can be done and how people live and use modern technology. Different industries such as automotive, finance, healthcare or manufacturing, can dramatically benefit from improved and faster data analysis, e.g., as illustrated by current industry trends like "Industry 4.0" and "Internet of Things". Data-driven research approaches utilizing Big Data technology and analysis become increasingly commonplace, e.g., in the life sciences, geo sciences or in astronomy. Users utilizing smartphones, social media, and web resources spend increasing amounts of time online, generate and consume enormous amounts of data and are the target for personalized services, recommendations and advertisements. Most of the possible developments related to Big Data are still in an early stage but there is great promise if the diverse technological and application-specific challenges in managing and using Big Data are successfully addressed. Some of the technical challenges have been associated to different "V" characteristics, in particular Volume, Velocity, Variety and Veracity that will also be discussed in this book. Other challenges relate to the protection of personal and sensitive data to ensure a high degree of privacy and the ability to turn the huge amount of data into useful insights or improved operation.

A key enabler for the Big Data movement are increasingly powerful and relatively inexpensive computing platforms allowing the fault-tolerant storage and processing of petabytes of data within large computing clusters typically equipped with thousands of processors and terabytes of main memory. The utilization of such infrastructures was pioneered by internet giants such as Google and Amazon but has become generally possible by open source system software such as the Hadoop ecosystem. Initially there have been only few core Hadoop components, in particular its distributed file system HDFS and the MapReduce framework for the relatively easy development and execution of highly parallel applications to process massive amounts of data on cluster infrastructures.

The initial Hadoop has been highly successful but also reached its limits in different areas, e.g., to support the processing of fast changing data such as data streams or to process highly iterative algorithms, e.g. for machine learning or graph processing. Furthermore, the Hadoop world has been largely decoupled from the widespread data management and analysis approaches based on relational databases and SQL. These aspects have led to a large number of additional components within the Hadoop ecosystem, both general-purpose processing frameworks such as Apache Spark and Flink as well as specific components, e.g., for data streams, graph data or machine learning. Furthermore, there are now numerous approaches to combine Hadoop-like data processing with relational database processing ("SQL on Hadoop").

The net effect of all these developments is that the current technological landscape for Big Data is not yet consolidated but there are many possible approaches within the Hadoop ecosystem but also within the product portfolio of different database vendors and other IT companies (Google, IBM, Microsoft, Oracle, etc.). The book "Big Data 2.0 Processing Systems" by Sherif Sakr is a valuable and up-to-date guide through this technological "jungle" and provides the reader with a comprehensible and concise overview of the main developments after the initial MapReduce-focused version of Hadoop. I am confident that this information is useful for many practitioners, scientists and students interested in Big Data technology.

Erhard Rahm

Preface

We live in an age of so-called *Big Data*. The radical expansion and integration of computation, networking, digital devices, and data storage has provided a robust platform for the explosion in big data as well as being the means by which big data are generated, processed, shared, and analyzed. In the field of computer science, data is considered as the main raw material which is produced by abstracting the world into categories, measures, and other representational forms (e.g., characters, numbers, relations, sounds, images, electronic waves) that constitute the building blocks from which information and knowledge are created. Big data has commonly been characterized by the defining 3Vs properties which refer to huge in *Volume*, consisting of terabytes or petabytes of data; high in *Velocity*, being created and processed in or near real-time; and diversity in *Variety* of type, being structured, semi-structured, and unstructured in nature. IDC predicts that the worldwide volume of data will reach 175 zettabytes by 2025, with a compounded annual growth rate of 61%, where 85% of all of this data will be residing in the cloud, as in data centers, and of new data types and formats including server logs and other machine-generated data, data from sensors, social media data, and many more other data sources. This new scale of Big Data has been attracting a lot of interest from both the research and industrial communities with the aim of creating the best means to process and analyze this data in order to make the best use of it. For about a decade, the MapReduce programming model and the Hadoop framework have dominated the world of big data processing; however, in recent years, academia and industry have started to recognize their limitations in several application domains and big data processing scenarios such as large-scale processing of structured data, graph data, streaming data, and machine/deep learning applications. Thus, the Hadoop framework is being actively replaced by a collection of engines which are dedicated to specific verticals (e.g., structured data, graph data, streaming data, machine learning, deep learning). In this book, we cover this new wave of systems referring to them as **Big Data 2.0 processing systems**.

This book provides a big picture and a comprehensive survey for the domain of big data processing systems. The book is not focused only on one research area or one type of data. However, it also discusses various aspects of research and development of big data systems. It also has a balanced descriptive and analytical content. It has information on advanced big data researches and also which parts of the research can benefit from further investigation. The book starts by introducing the general background of the big data phenomena. We then provide an overview of various general-purpose big data processing systems which empower its user to develop various big data processing jobs for different application domains. We next examine the several vertical domains of big data processing systems: structured data, graph data, stream data, and machine/deep learning applications. This book is concluded with a discussion for some of the open problems and future research directions.

We hope this monograph could be a useful reference for students, researchers, and professionals in the domain of big data processing systems. We also wish that the comprehensive reading materials of the book could influence readers to think further and investigate the areas that are novel to them.

To Students: We hope that the book provides you with an enjoyable introduction to the field of big data processing systems. We have attempted to properly classify the state of the art and describe technical problems and techniques/methods in depth. The book will provide you with a comprehensive list of potential research topics. You can use this book as a fundamental starting point for your literature survey.

To Researchers: The material of this book will provide you with a thorough coverage for the emerging and ongoing advancements on big data processing systems which are being designed to deal with specific verticals in addition to the general-purpose ones. You can use the chapters that are related to their research interest as a solid literature survey. You also can use this book as a starting point to endeavor other research topics.

To Professionals and Practitioners: You will find this book useful as it provides a review of the state of the art for big data processing systems. The wide range of systems and techniques covered in this book makes it an excellent handbook on big data analytics systems. Most of the problems and systems that we discuss in each chapter have great practical utility in various application domains. The reader can immediately put the gained knowledge from this book into practice due to the open-source availability of the majority of the big data processing systems.

Tartu, Estonia   Sherif Sakr

Acknowledgements

I am grateful to many of my collaborators for their contribution to this book. In particular, I would like to mention, Fuad Bajaber, Ahmed Barnawi, Omar Batarfi, Seyed-Reza Beheshti, Radwa Elshawi, Ayman Fayoumi, Anna Liu, and Reza Nouri. Thank you all!

Thanks to Springer-Verlag for publishing this book. Ralf Gerstner encouraged and supported me to write this book. Thanks Ralf!

My acknowledgements end with thanking the people most precious. Thanks to my parents for their encouragement and support. Many thanks to my daughter, Jana and my son, Shehab for the happiness and enjoyable moments they are always bringing to my life. My most special appreciation goes to my wife, Radwa Elshawi, for her everlasting support and deep love.

Tartu, Estonia   Sherif Sakr

Contents

1 Introduction 1
   1.1   The Big Data Phenomena 1
   1.2   Big Data and Cloud Computing 5
   1.3   Big Data Storage Systems... 7
   1.4   Big Data Processing and Analytics Systems 11
   1.5   Book Road Map 14

2 General-Purpose Big Data Processing Systems.......................... 17
   2.1   The Big Data Star: The Hadoop Framework 17
        2.1.1   The Original Architecture...................................... 17
        2.1.2   Enhancements of the MapReduce Framework 21
        2.1.3   Hadoop's Ecosystem 29
   2.2   Spark 30
   2.3   Flink 36
   2.4   Hyracks/ASTERIX 40

3 Large-Scale Processing Systems of Structured Data 45
   3.1   Why SQL-On-Hadoop? 45
   3.2   Hive 46
   3.3   Impala 49
   3.4   IBM Big SQL 50
   3.5   SPARK SQL 51
   3.6   HadoopDB 53
   3.7   Presto 54
   3.8   Tajo 56
   3.9   Google Big Query 57
   3.10 Phoenix... 57
   3.11 Polybase 58

4 Large-Scale Graph Processing Systems 59
   4.1   The Challenges of Big Graphs 59
   4.2   Does Hadoop Work Well for Big Graphs? 61
   4.3   Pregel Family of Systems 64
        4.3.1   The Original Architecture...................................... 64
        4.3.2   Giraph: BSP + Hadoop for Graph Processing 67
        4.3.3   Pregel Extensions... 70

	4.4	GraphLab Family of Systems	72
	4.4.1	GraphLab	72
	4.4.2	PowerGraph	73
	4.4.3	GraphChi	73
	4.5	Spark-Based Large-Scale Graph Processing Systems	75
	4.6	Gradoop	77
	4.7	Other Systems	79
	4.8	Large-Scale RDF Processing Systems	81
	4.8.1	NoSQL-Based RDF Systems	82
	4.8.2	Hadoop-Based RDF Systems	85
	4.8.3	Spark-Based RDF Systems	87
	4.8.4	Other Distributed RDF Systems	89

5   Large-Scale Stream Processing Systems 95
    5.1   The Big Data Streaming Problem 95
    5.2   Hadoop for Big Streams?! 98
    5.3   Storm 101
    5.4   Infosphere Streams 103
    5.5   Other Big Stream Processing Systems 105
    5.6   Big Data Pipelining Frameworks 110
        5.6.1 Pig Latin 110
        5.6.2 Tez 112
        5.6.3 Other Pipelining Systems 114

6   Large-Scale Machine/Deep Learning Frameworks 117
    6.1   Harnessing the Value of Big Data 117
    6.2   Big Machine Learning Frameworks 118
    6.3   Deep Learning Frameworks 123

7   Conclusions and Outlook 127

References 135

About the Author

Prof. Sherif Sakr passed away on Thursday, 25 March 2020 at the age of 40. He was the Head of Data Systems Group at the Institute of Computer Science, University of Tartu. He received his PhD degree in Computer and Information Science from Konstanz University, Germany in 2007. He received his BSc and MSc degrees in Computer Science from the Information Systems department at the Faculty of Computers and Information in Cairo University, Egypt, in 2000 and 2003, respectively. Before joining University of Tartu, he held appointments in several international organizations including University of New South Wales (Australia), Macquarie University (Australia), Data61/CSIRO (Australia), Microsoft Research (USA), Nokia Bell Labs (Ireland), and King Saud Bin Abdulaziz University for Health Sciences (Saudi Arabia). He also held several visiting appointments at Humboldt-Universitat zu Berlin (Germany), University of Zurich (Switzerland), and Technical University of Dresden (Germany).

Prof. Sakr's research interest is data and information management in general, particularly in big data processing systems, big data analytics, data science, and big data management in cloud computing platforms. He published more than 150 refereed research publications in international journals and conferences. He was an ACM Senior Member and an IEEE Senior Member. In 2017, he had been appointed to serve as an ACM Distinguished Speaker and as an IEEE Distinguished Speaker. He served as the Editor in Chief of the Springer Encyclopedia of Big Data Technologies.

He was also serving as a Co-Chair for the European Big Data Value Association (BDVA) TF6-Data Technology Architectures Group. In 2016, he received the best research award from King Abdullah International Medical Research Center (KAIMRC). In 2019, he received the best Arab scholar award from Abdul Hameed Shoman Foundation.

Introduction

1

1.1   The Big Data Phenomena

There is no doubt that we are living the era of big data where we are witnessing radical expansion and integration of digital devices, networking, data storage, and computation systems. In practice, data generation and consumption is becoming a main part of people's daily life especially with the pervasive availability and usage of Internet technology and applications [1]. The number of Internet users reached 4.5 billion in 2019.[1] As a result, we are witnessing an explosion in the volume of creation of digital data from various sources and at ever-increasing rates. Social networks, mobile applications, cloud computing, sensor networks, video surveillance, GPS, RFID, Internet of Things (IoT), imaging technologies, and gene sequencing are just examples of technologies that facilitate and accelerate the continuous creation of massive datasets that must be stored and processed. For example, in 1 min on the Internet,[2] we have 1 million Facebook logins, 4.5 million videos watched on YouTube, 1.4 million swipes on Tinder (right or left, we cannot say for sure), and a total of 41.6 million messages sent on WhatsApp and Facebook Messenger. That same internet minute also contains 3.8 million Google queries, 347,222 scrolls on Instagram, and almost a million dollars spent online (Fig. 1.1). These numbers, which are continuously increasing, provide a perception of the massive data generation, consumption, and traffic which are happening in the Internet world. In another context, powerful telescopes in astronomy, particle accelerators in physics, and genome sequencers in biology are producing vast volumes of data into the hands of scientists. The cost of sequencing one human genome has fallen from $100 million in 2001 to around $1K in 2019[3] (Fig. 1.2).

---

[1] https://www.internetworldstats.com/stats.htm.

[2] https://www.statista.com/chart/17518/internet-use-one-minute/.

[3] https://www.genome.gov/about-genomics/fact-sheets/DNA-Sequencing-Costs-Data.

© The Editor(s) (if applicable) and The Author(s), under exclusive
licence to Springer Nature Switzerland AG 2020
S. Sakr, *Big Data 2.0 Processing Systems*,
https://doi.org/10.1007/978-3-030-44187-6_1

A minute on the Internet in 2019

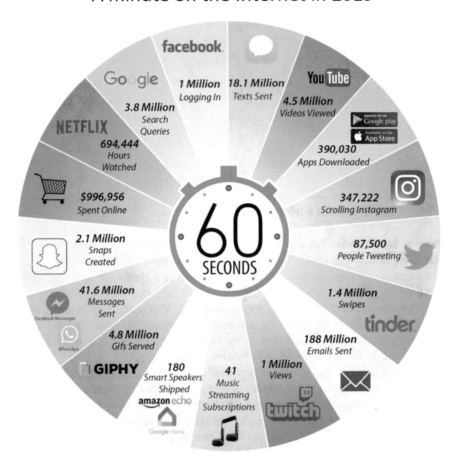

Fig. 1.1 One minute in the Internet

Every day, Survey Telescope [2] generates on the order of 30 TeraBytes of data on a daily basis, the New York Stock Exchange captures around 1 TB of trade information, and about 30 billion radio-frequency identification (RFID) tags are created. Add to this mix the data generated by the hundreds of millions of GPS devices sold every year, and the more than 30 million networked sensors currently in use (and growing at a rate faster than 30% per year). These data volumes are expected to double every 2 years over the next decade. IBM reported that we are

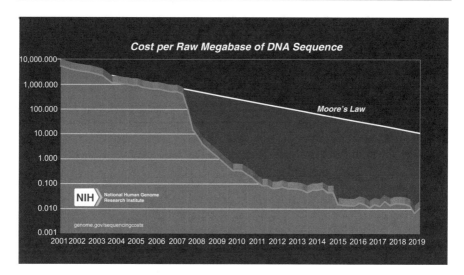

Fig. 1.2 Genome sequencing cost

currently producing 2.5 quintillion bytes of data everyday.[4] IDC predicts that the worldwide volume of data will reach 175 zettabytes by 2025[5] where 85% of all of this data will be of new data types and formats including server logs and other machine generated data, data from sensors, social media data, and many more other data sources (Fig. 1.3). All of this data will enable us as to do things which we were not able to before and thereby create value for the world economy. However, clearly, many application domains are facing major challenges on processing such massive amount of generated data from different sources and in various formats. Therefore, almost all scientific funding and government agencies introduced major strategies and plans to support big data research and applications.

In the enterprise world, many companies continuously gather massive datasets that store customer interactions, product sales, results from advertising campaigns on the Web in addition to various types of other information [3]. In practice, a company can generate up to petabytes of information over the course of a year: web pages, clickstreams, blogs, social media forums, search indices, email, documents, instant messages, text messages, consumer demographics, sensor data from active and passive systems, and more. By many estimates, as much as 80% of this data is semi-structured or unstructured. In practice, it is typical that companies are always seeking to become more nimble in their operations and more innovative with their data analysis and decision-making processes. They are realizing that time lost in these processes can lead to missed business opportunities. The core of the data

---

[4]http://www-01.ibm.com/software/data/bigdata/what-is-big-data.html.

[5]https://www.networkworld.com/article/3325397/idc-expect-175-zettabytes-of-data-worldwide-by-2025.html.

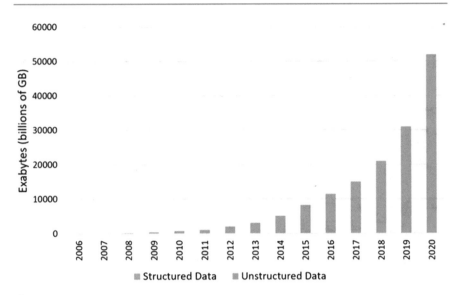

Fig. 1.3 Digital data growth

management challenge is for companies to gain the ability to analyze and understand Internet-scale information just as easily as they can now analyze and understand smaller volumes of structured information.

The *Big Data* term has been coined under the tremendous and explosive growth of the world digital data which is generated from various sources and in different formats. In principle, the Big Data term is commonly described by 3V main attributes (Fig. 1.4): the *Volume* attributes describe the massive amount of data that can be billions of rows and millions of columns, the *Variety* attribute represents the variety of formats, data sources, and structures, and the *Velocity* attribute reflects the very high speed on data generation, ingestion, and near real-time analysis. In January 2007, Jim Gray, a database scholar, described the big data phenomena as the *Fourth Paradigm* [4] and called for a *paradigm shift* in the computing architecture and large-scale data processing mechanisms. The first three paradigms were *experimental, theoretical* and, more recently, *computational science*. Gray argued that the only way to cope with this paradigm is to develop a new generation of computing tools to manage, visualize, and analyze the data flood. According to Gray, computer architectures have become increasingly imbalanced where the latency gap between multicore CPUs and mechanical hard disks is growing every year which makes the challenges of data-intensive computing much harder to overcome [5]. Hence, there is a crucial need for a systematic and generic approach to tackle these problems with an architecture that can also scale into the foreseeable future. In response, Gray argued that the new trend should instead focus on supporting cheaper clusters of computers to manage and process all this data instead of focusing on having the biggest and fastest single computer. In addition, the 2011 McKinsey global report described big data as the next frontier for innovation and

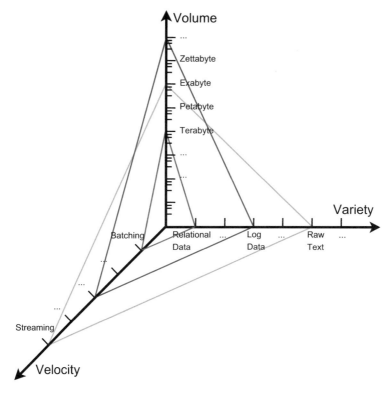

Fig. 1.4 3V characteristics of big data

competition [6]. The report defined big data as "*Data whose scale, distribution, diversity, and/or timeliness require the use of new technical architectures and analytics to enable insights that unlock the new sources of business value.*" This definition highlighted the crucial need for new data architecture solution that can manage the increasing challenges of big data problems. In response, the new scale of *Big Data* has been attracting a lot of interest from both the research and industrial worlds aiming to create the best means to process and analyze this data and make the best use of it [7].

1.2 Big Data and Cloud Computing

Over the last decade, there has been a great deal of hype about cloud computing [8]. In principle, cloud computing is associated with a new paradigm for the provision of computing infrastructure. This paradigm shifts the location of this infrastructure to more centralized and larger scale data centers in order to reduce the costs associated with the management of hardware and software resources (Fig. 1.5).

Fig. 1.5 Cloud computing

In fact, the discussion in industry and academia has taken a while until they are able to define the road map for defining what cloud computing actually means [9–11]. The US National Institute of Standards and Technology (NIST) has published a definition that reflects the most commonly agreed features of cloud computing. This definition describes the cloud computing technology as: "*a model for enabling convenient, on-demand network access to a shared pool of configurable computing resources (e.g., networks, servers, storage, applications, and services) that can be rapidly provisioned and released with minimal management effort or service provider interaction.*" In principle, one of the important features provided by the cloud computing technology is that computing hardware and software capabilities are made accessible via the network and accessed through standard mechanisms that can be supported by heterogeneous thin or fat client platforms (e.g., laptops, mobile phones, and PDAs). In particular, cloud computing provided a number of advantages for hosting the deployments of data-intensive applications such as:

- Reduced time-to-market by removing or simplifying the time-consuming hardware provisioning, purchasing, and deployment processes.
- Reduced monetary cost by following a *pay-as-you-go* business model.
- Unlimited (virtually) computing resources and scalability by adding resources as the workload increases.

Therefore, cloud computing has been considered as a significant step toward achieving the long-held dream of envisioning computing as a utility [12] where the economies of scale principles help to drive the cost of computing infrastructure effectively down. In practice, big players of the technology companies (e.g., Amazon, Microsoft, Google, IBM, Oracle) have been quite active on establishing their own data centers across the world to ensure reliability by providing redundancy for their provided infrastructure, platforms, and applications to the end users. In principle, cloud-based services offer several advantages such as flexibility and scalability of storage, computing and application resources, optimal utilization of infrastructure, and reduced costs. Hence, cloud computing provides a great chance to supply storage, processing, and analytics resources for big data applications. A recent analysis[6] reported that 53% of enterprises have deployed (28%) or plan to deploy (25%) their big data analytics (BDA) applications in the Cloud.

In cloud computing, the provider's computing resources are pooled to serve multiple consumers using a multitenant model with various virtual and physical resources dynamically assigned and reassigned based on the demand of the application workload. Therefore, it achieves the sense of location independence. Examples of such shared computing resources include storage, memory, network bandwidth, processing, virtual networks, and virtual machines. In practice, one of the main principles for the data centers technology is to exploit the virtualization technology to increase the utilization of computing resources. Hence, it supplies the main ingredients of computing resources such as CPUs, storage, and network bandwidth as a commodity at low unit cost. Therefore, users of cloud services do not need to be concerned about the problem of resource scalability because the provided resources can be virtually considered as being infinite. In particular, the business model of public cloud providers relies on the mass-acquisition of IT resources which are made available to cloud consumers via various attractive pricing models and leasing packages. This provides applications or enterprises with the opportunity to gain access to powerful infrastructure without the need to purchase it.

1.3 Big Data Storage Systems

In general, relational database management systems (e.g., MySQL, PostgreSQL, SQL Server, Oracle) have been considered as the *one-size-fits-all* solution for data persistence and retrieval for decades. They have matured after extensive research and development efforts and very successfully created a large market and many solutions in different business domains. However, the ever-increasing need for scalability and new application requirements have created new challenges for traditional RDBMS [13]. In particular, we are currently witnessing a continuous increase of user-driven and user-generated data that resulted in a tremendous growth in the type and volume of data which is produced, stored, and analyzed. For example, various newer sets of sources for data sources are emerging such as: sensor technologies, automated trackers, Global Positioning Systems (GPS) and monitoring devices are producing massive datasets. In addition to the speedy data

---

[6]https://www.thorntech.com/2018/09/big-data-in-the-cloud/.

growth, data has also become increasingly of sparse and semi-structured in nature. In particular, data structures can be classified into four main types as follows:

- **Structured data**: Data which has a defined format and structure such as CSV files, spreadsheets, traditional relational databases, and OLAP data cubes.
- **Semi-structured data**: Textual data files with a flexible structure that can be parsed. The popular example of such type of data is the Extensible Markup Language (XML) data files with its self-describing information.
- **Quasi-structured data**: Textual data with erratic data formats such as web clickstream data that may contain inconsistencies in data values and formats.
- **Unstructured data**: Data that has no inherent structure such as text documents, images, PDF files, and videos.

Figure 1.6 illustrates the data structure evolution over the years. In practice, the continuous growth in the sizes of such types of data led to the challenge that the traditional data management techniques that required upfront schema definition and relational-based data organization are inadequate in many scenarios. Therefore, in order to tackle this challenge, we have witnessed the emergence of a new generation of scalable data storage system called *NoSQL* (**N**ot **O**nly **SQL**) database systems. This new class of database systems can be classified into four main types (Fig. 1.7):

- *Key-value stores*: These systems use the simplest data model which is a collection of objects where each object has a unique key and a set of attribute/value pairs.

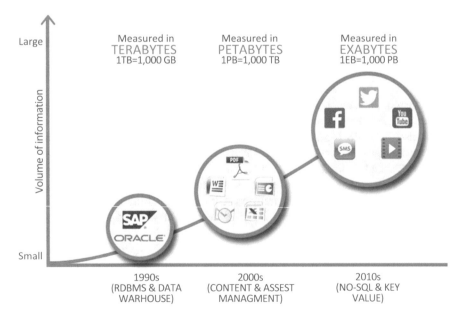

Fig. 1.6 The data structure evolution over the years

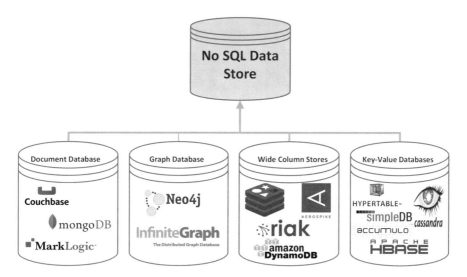

Fig. 1.7 Types of NoSQL data stores

- *Extensible record stores*: They provide variable-width tables (column families) that can be partitioned vertically and horizontally across multiple servers.
- *Document stores*: The data model of these systems consists of objects with a variable number of attributes with a possibility of having nested objects.
- *Graph stores*: The data model of these systems uses graph structures with edges, nodes, and properties to model and store data.

In general, scalability represents the capability of a system to increase throughput via increasing the allocated resources to handle the increasing workloads [14]. In practice, scalability is usually accomplished either by provisioning additional resources to meet the increasing demands (vertical scalability) or it can be accomplished by grouping a cluster of commodity machines to act as an integrated work unit (horizontal scalability). Figure 1.8 illustrates the comparison between the horizontal and vertical scalability schemes. In principle, vertical scaling option is typically expensive and proprietary, while horizontal scaling is achieved by adding more nodes to manage additional workloads which fits well with the *pay-as-you-go* pricing philosophy of the emerging cloud computing models. In addition, vertical scalability normally faces an absolute limit that cannot be exceeded, no matter how much resources can be added or how much money one can spend. Furthermore, horizontal scalability leads to the fact that the storage system would become more resilient to fluctuations in the workload because handling of separate requests is managed such that they do not have to compete on shared hardware resources.

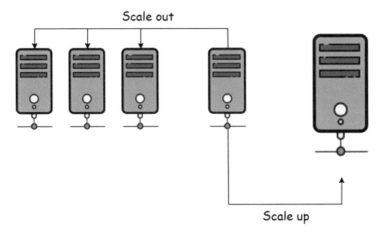

Fig. 1.8 Horizontal scalability vs vertical scalability

In practice, many systems[7] that are identified to fall under the umbrella of NoSQL systems are quite varied; however, each of these systems comes with their unique sets of features and value propositions [15]. For example, the key/value (KV) data stores represent the simplest model of NoSQL systems which pairs keys to values in a very similar fashion to how a map (or hash table) works in any standard programming language. Various open-source projects have been implemented to provide key-value NoSQL database systems such as *Memcached*,[8] *Voldemort*,[9] *Redis*,[10] and *Riak*.[11] Columnar, or column-oriented, is another type of NoSQL databases. In such systems, data from a given column is stored together in contrast to a row-oriented database (e.g., relational database systems) which keeps information about each row together. In column-oriented databases, adding new columns is quite flexible and is performed on the fly on a row-by-row basis. In particular, every row may have a different set of columns which allow tables to be sparse without introducing any additional storage cost for null values. In principle, columnar NoSQL systems represent a midway between relational and key-value stores. *Apache HBase*[12] is currently the most popular open-source system of this category. Another category of NoSQL systems is document-oriented database stores. In this category, a document is like a hash, with a unique ID field and values that may be any of a variety of types, including more hashes. In particular, documents can contain nested structures, and so they provide a high

---

[7]http://nosql-database.org/.

[8]http://memcached.org/.

[9]http://www.project-voldemort.com/voldemort/.

[10]http://redis.io/.

[11]http://basho.com/riak/.

[12]http://hbase.apache.org/.

degree of flexibility, allowing for variable domains. *MongoDB*[13] and *CouchDB*[14] are currently the two most popular systems in this category. Finally, NoSQL Graph databases is another category which excel in handling highly interconnected data. In principle, a graph database consists of nodes and relationships between nodes where both relationships and nodes can be described using descriptive information and properties (key-value pairs). In principle, the main advantage of graph databases is that they provide easy functionalities for traversing through the nodes of the graph structure by following relationships. The *Neo4J*[15] database system is currently the most popular in this category.

1.4 Big Data Processing and Analytics Systems

There is no doubt that our societies have become increasingly more instrumented in which we are producing and storing vast amounts of data. As a result, in our modern world, data is key resource. However, in practice, data are not useful in and of themselves. They only have utility if meaning and value can be extracted from them. Therefore, given their utility and value, there are always continuous increasing efforts devoted to producing and analyzing them. In principle, big data discovery enables data scientists and other analysts to uncover patterns and correlations through analysis of large volumes of data of diverse types. In particular, the power of big data is revolutionizing the way we live. From the modern business enterprise to the lifestyle choices of today's digital citizen, the insights of big data analytics are driving changes and improvements in every arena [16]. For instance, insights gleaned from big data discovery can provide businesses with significant competitive advantages, such as more successful marketing campaigns, decreased customer churn, and reduced loss from fraud. In particular, these insights provide the opportunity to make businesses more agile and answer queries that were previously considered beyond their reach. Therefore, it is crucial that all the emerging varieties of data types with huge sizes need to be harnessed to provide a more complete picture of what is happening in various application domains. In particular, in the current era, data represent the new gold, while analytics systems represent the machinery that analyzes, mines, models, and mints it.

In practice, the increasing demand for large-scale data analysis and data mining applications has stimulated designing and building novel solutions from both the industry (e.g., click-stream analysis, web-data analysis, network-monitoring log analysis) and the sciences (e.g., analysis of data produced by massive-scale simulations, sensor deployments, high-throughput lab equipment) [17]. Although parallel database systems [18] serve some of these data analysis applications (e.g.,

---

[13]http://www.mongodb.org/.

[14]http://couchdb.apache.org/.

[15]http://neo4j.com/.

Teradata,[16] SQL Server PDW,[17] Vertica,[18] Greenplum,[19] ParAccel,[20] Netezza,[21]) they are expensive, difficult to administer, and lack fault tolerance for long-running queries [19].

In 2004, Google made a seminal contribution to the big data world by introducing the MapReduce framework as a simple and powerful programming model that enables easy development of scalable parallel applications to process vast amounts of data on large clusters of commodity machines by scanning and processing large files in parallel across multiple machines [20]. In particular, the framework is mainly designed to achieve high performance on large clusters of commodity PCs. The fundamental principle of the MapReduce framework is to move analysis to the data, rather than moving the data to a system that can analyze it. One of the main advantages of this approach is that it isolates the application from the details of running a distributed program, such as issues on data distribution, scheduling, and fault tolerance. Thus, it allows programmers to think in a *data-centric* fashion where they can focus on applying transformations to sets of data records while the details of distributed execution and fault tolerance are transparently managed by the MapReduce framework.

In practice, the Hadoop project,[22] the open-source realization of the MapReduce framework, has been a big success and created an increasing momentum in the research and business domains. In addition, cost-effective processing of large datasets is a nontrivial undertaking. Fortunately, MapReduce frameworks and cloud computing have made it easier than ever for everyone to step into the world of big data. This technology combination has enabled even small companies to collect and analyze terabytes of data in order to gain a competitive edge [21]. For example, the Amazon Elastic Compute Cloud (EC2)[23] is offered as a commodity that can be purchased and utilized. In addition, Amazon has also provided the Amazon Elastic MapReduce[24] as an online service to easily and cost-effectively process vast amounts of data without the need to worry about time-consuming setup, management or tuning of computing clusters, or the compute capacity upon which they sit. Hence, such services enable third parties to perform their analytical queries on massive datasets with minimum effort and cost by abstracting the complexity entailed in building and maintaining computer clusters. Therefore, due to its success, it has been supported by many big players in their big data commercial platforms

---

[16]http://teradata.com/.

[17]http://www.microsoft.com/sqlserver/en/us/solutions-technologies/data-warehousing/pdw.aspx.

[18]http://www.vertica.com/.

[19]http://www.greenplum.com/.

[20]http://www.paraccel.com/.

[21]http://www-01.ibm.com/software/data/netezza/.

[22]http://hadoop.apache.org/.

[23]http://aws.amazon.com/ec2/.

[24]http://aws.amazon.com/elasticmapreduce/.

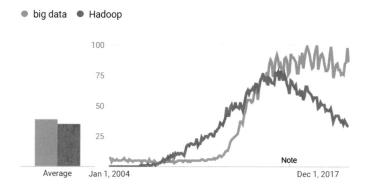

Fig. 1.9 Google's Web search trends for the two search items: Big Data and Hadoop (created by Google Trends)

such as *Microsoft*,[25] *IBM*,[26] and *Oracle*.[27] In addition, several successful startups such as *MapR*,[28] *Cloudera*,[29] *Altiscale*,[30] *Splice Machine*,[31] *DataStax*,[32] *Platfora*,[33] and *Trifacta*[34] have built their solutions and services based on the Hadoop project. Figure 1.9 illustrates Google's web search trends for the two search items: *Big Data* and *Hadoop*, according to the Google trend analysis tool.[35] In principle, Fig. 1.9 shows that the search item *Hadoop* has overtaken the search item *Big Data* and has since dominated the Web users' search requests during the period between 2008 and 2012 while since 2013, the two search items have started to go side by side.

Recently, both the research and industrial domains have identified various limitations in the Hadoop framework [22] and thus there is now common consensus that the Hadoop framework cannot be the *one-size-fits-all* solution for the various big data analytics challenges. Therefore, in this book, we argue that the Hadoop framework with its extensions [22] represented the Big Data 1.0 processing platforms. We coin the term of Big Data 2.0 processing platforms which represent a new generation of engines that are domain-specific, dedicated to specific verticals,

---

[25]http://azure.microsoft.com/en-us/services/hdinsight/.

[26]http://www-01.ibm.com/software/data/infosphere/hadoop/enterprise.html.

[27]http://www.oracle.com/us/products/middleware/data-integration/hadoop/overview/index.html.

[28]https://www.mapr.com/.

[29]http://www.cloudera.com/.

[30]https://www.altiscale.com/.

[31]http://www.splicemachine.com/.

[32]http://www.datastax.com/.

[33]https://www.platfora.com/.

[34]http://www.trifacta.com/.

[35]http://www.google.com/trends/.

Fig. 1.10 Timeline representation of Big Data 2.0 processing platforms. Flags denote the general-purpose big data processing systems. Rectangles denote the big SQL processing platforms, stars denote large-scale graph processing platforms, and diamonds denote large-scale stream processing platforms

and slowly replacing the Hadoop framework in various usage scenarios. Figure 1.10 illustrates a timeline view for the development of the Big Data 2.0 processing platforms. Notably, there has been growing activities around the big data hotspot at the academic and industrial worlds, mainly from 2009 and onwards, focused on building a new generation of optimized and domain-specific big data analytics platforms. The main focus of this book is to highlight and provide an overview of this new generation of systems.

1.5   Book Road Map

Figure 1.11 illustrates a classification of these emerging systems which we are going to detail in the next. In general, the discovery process often employs analytics techniques from a variety of genres such as time-series analysis, text analytics, statistics, and machine learning. Moreover, the process might involve the analysis of structured data from conventional transactional sources, in conjunction with the analysis of multi-structured data from other sources such as clickstreams, call detail records, application logs, or text from call center records. Chapter 2 provides an overview of various general-purpose big data processing systems which empower its user to develop various big data processing jobs for different application domains.

Several studies reported that Hadoop is not an adequate choice for supporting interactive queries which aim of achieving a response time of milliseconds or few seconds [19]. In addition, many programmers may be unfamiliar with the Hadoop framework and they would prefer to use SQL as a high-level declarative language to implement their jobs while delegating all of the optimization details in the execution process to the underlying engine [22]. Chapter 3 provides an overview of various systems that have been introduced to support the SQL flavor on top of the Hadoop infrastructure and provide competing and scalable performance on processing large-scale structured data.

Fig. 1.11 Classification of Big Data 2.0 processing systems

Nowadays, graphs with millions and billions of nodes and edges have become very common. The enormous growth in graph sizes requires huge amounts of computational power to analyze. In general, graph processing algorithms are iterative and need to traverse the graph in a certain way. Chapter 4 focuses on discussing several systems that have been designed to tackle the problem of large-scale graph processing.

In general, stream computing is a new paradigm which has been necessitated by new data-generating scenarios, such as the ubiquity of mobile devices, location services, and sensor pervasiveness. In general, stream processing engines enable a large class of applications in which data are produced from various sources and are moved asynchronously to processing nodes. Thus, streaming applications are normally configured as continuous tasks in which their execution starts from the time of their inception till the time of their cancellation. The main focus of Chap. 5 is to cover several systems that have been designed to provide scalable solutions for processing big data streams in addition to other set of systems that have been introduced to support the development of data pipelines between various types of big data processing jobs and systems.

With the wide availability of data and increasing capacity of computing resources, machine learning and deep learning techniques have become very popular techniques on harnessing the power of data by achieving powerful analytical features. Chapter 6 focuses on discussing several systems that have been developed to support computationally expensive machine learning and deep learning algorithms on top of big data processing frameworks. Finally, we provide some conclusions and outlook for future research challenges in Chap. 7.

General-Purpose Big Data Processing Systems

2

2.1 The Big Data Star: The Hadoop Framework

2.1.1 The Original Architecture

In 2004, Google has introduced the MapReduce framework as a simple and powerful programming model that enables easy development of scalable parallel applications to process vast amounts of data on large clusters of commodity machines [20]. In particular, the implementation described in the original paper is mainly designed to achieve high performance on large clusters of commodity PCs. One of the main advantages of this approach is that it isolates the application from the details of running a distributed program, such as issues on data distribution, scheduling, and fault tolerance. In this model, the computation takes a set of key/value pairs as input and produces a set of key/value pairs as output. The user of the MapReduce framework expresses the computation using two functions: *Map* and *Reduce*. The Map function takes an input pair and produces a set of intermediate key/value pairs. The MapReduce framework groups together all intermediate values associated with the same intermediate key I and passes them to the Reduce function. The Reduce function receives an intermediate key I with its set of values and merges them together. Typically just zero or one output value is produced per Reduce invocation. The main advantage of this model is that it allows large computations to be easily parallelized and re-executed to be used as the primary mechanism for fault tolerance. Figure 2.1 illustrates an example MapReduce program expressed in pseudo-code for counting the number of occurrences of each word in a collection of documents. In this example, the Map function emits each word plus an associated count of occurrences while the Reduce function sums together all counts emitted for

S. Sakr, *Big Data 2.0 Processing Systems*, https://doi.org/10.1007/978-3-030-44187-6_2

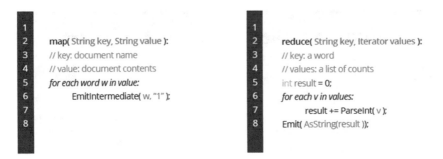

```
1
2    map( String key, String value ):
3    // key: document name
4    // value: document contents
5    for each word w in value:
6         EmitIntermediate( w, "1" );
7
8
```

```
1
2    reduce( String key, Iterator values ):
3    // key: a word
4    // values: a list of counts
5    int result = 0;
6    for each v in values:
7         result += ParseInt( v );
8    Emit( AsString(result ));
```

Fig. 2.1   An example MapReduce program

a particular word. In principle, the design of the MapReduce framework is based on the following main principles [23]:

- *Low-Cost Unreliable Commodity Hardware*: Instead of using expensive, high performance, reliable symmetric multiprocessing (SMP) or massively parallel processing (MPP) machines equipped with high-end network and storage sub-systems, the MapReduce framework is designed to run on large clusters of commodity hardware. This hardware is managed and powered by open-source operating systems and utilities so that the cost is low.
- *Extremely Scalable RAIN Cluster*: Instead of using centralized RAID-based SAN or NAS storage systems, every MapReduce node has its own local off-the-shelf hard drives. These nodes are loosely coupled where they are placed in racks that can be connected with standard networking hardware connections. These nodes can be taken out of service with almost no impact to still-running MapReduce jobs. These clusters are called redundant array of independent (and inexpensive) nodes (RAIN).
- *Fault-Tolerant Yet Easy to Administer*: MapReduce jobs can run on clusters with thousands of nodes or even more. These nodes are not very reliable as at any point in time, a certain percentage of these commodity nodes or hard drives will be out of order. Hence, the MapReduce framework applies straightforward mechanisms to replicate data and launch backup tasks so as to keep still-running processes going. To handle crashed nodes, system administrators simply take crashed hardware offline. New nodes can be plugged in at any time without much administrative hassle. There are no complicated backup, restore, and recovery configurations like the ones that can be seen in many DBMSs.
- *Highly Parallel Yet Abstracted*: The most important contribution of the MapReduce framework is its ability to automatically support the parallelization of task executions. Hence, it allows developers to focus mainly on the problem at hand rather than worrying about the low-level implementation details such as memory management, file allocation, parallel, multi-threaded or network programming. Moreover, MapReduce's shared-nothing architecture [24] makes it much more scalable and ready for parallelization.

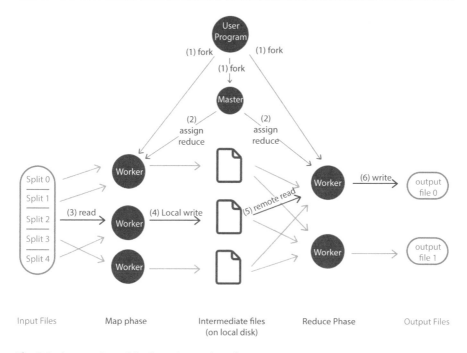

Fig. 2.2   An overview of the flow of execution of a MapReduce operation

Hadoop[1] is an open-source Java library [25] that supports data-intensive distributed applications by realizing the implementation of the MapReduce framework.[2] It has been widely used by a large number of business companies for production purposes.[3] On the implementation level, the Map invocations of a MapReduce job are distributed across multiple machines by automatically partitioning the input data into a set of M splits. The input splits can be processed in parallel by different machines. Reduce invocations are distributed by partitioning the intermediate key space into R pieces using a partitioning function (e.g., hash(key) mod R). The number of partitions (R) and the partitioning function are specified by the user. Figure 2.2 illustrates an example of the overall flow of a MapReduce operation which goes through the following sequence of actions:

1. The input data of the MapReduce program is split into M pieces and starts up many instances of the program on a cluster of machines.
2. One of the instances of the program is elected to be the *master* copy while the rest are considered as *workers* that are assigned their work by the master copy. In

---

[1]http://hadoop.apache.org/.

[2]In the rest of this article, we use the two names: MapReduce and Hadoop, interchangeably.

[3]http://wiki.apache.org/hadoop/PoweredBy.

particular, there are M map tasks and R reduce tasks to assign. The master picks idle workers and assigns each one or more map tasks and/or reduce tasks.

3. A worker who is assigned a map task processes the contents of the corresponding input split and generates key/value pairs from the input data and passes each pair to the user-defined Map function. The intermediate key/value pairs produced by the Map function are buffered in memory.

4. Periodically, the buffered pairs are written to local disk and partitioned into R regions by the partitioning function. The locations of these buffered pairs on the local disk are passed back to the master, who is responsible for forwarding these locations to the reduce workers.

5. When a reduce worker is notified by the master about these locations, it reads the buffered data from the local disks of the map workers which is then sorted by the intermediate keys so that all occurrences of the same key are grouped together. The sorting operation is needed because typically many different keys map to the same reduce task.

6. The reduce worker passes the key and the corresponding set of intermediate values to the user's Reduce function. The output of the Reduce function is appended to a final output file for this reduce partition.

7. When all map tasks and reduce tasks have been completed, the master program wakes up the user program. At this point, the MapReduce invocation in the user program returns the program control back to the user code.

Figure 2.3 illustrates a sample execution for the example program ($\mathtt{WordCount}$) depicted in Fig. 2.1 using the steps of the MapReduce framework which are illustrated in Fig. 2.2. During the execution process, the master pings every worker

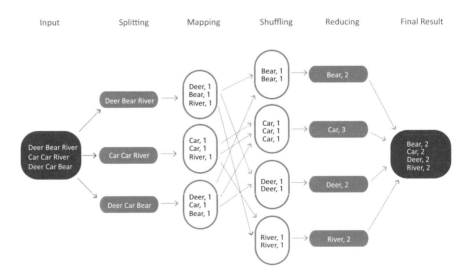

Fig. 2.3 Execution steps of the word count example using the MapReduce framework

periodically. If no response is received from a worker within a certain amount of time, the master marks the worker as *failed*. Any map tasks marked *completed* or *in progress* by the worker are reset back to their initial idle state and therefore become eligible for scheduling by other workers. Completed map tasks are re-executed on a task failure because their output is stored on the local disk(s) of the failed machine and is therefore inaccessible. Completed reduce tasks do not need to be re-executed since their output is stored in a global file system.

The Hadoop project has been introduced as an open-source Java library that supports data-intensive distributed applications and clones the implementation of the Google's MapReduce framework [20]. In principle, Hadoop framework consists of two main components: the Hadoop Distributed File System (HDFS) and the MapReduce programming model. In particular, HDFS provides the basis for distributed big data storage which distributes the data files into data blocks and stores such data in different nodes of the underlying computing cluster in order to enable effective parallel data processing.

2.1.2   Enhancements of the MapReduce Framework

In practice, the basic implementation of the MapReduce is very useful for handling data processing and data loading in a heterogeneous system with many different storage systems. Moreover, it provides a flexible framework for the execution of more complicated functions than that can be directly supported in SQL. However, this basic architecture suffered from some limitations. In the following subsections we discuss some research efforts that have been conducted in order to deal with these limitations by providing various enhancements on the basic implementation of the MapReduce framework.

2.1.2.1 Processing Join Operations
One main limitation of the MapReduce framework is that it does not support the joining of multiple datasets in one task. However, this can still be achieved with additional MapReduce steps. For example, users can map and reduce one dataset and read data from other datasets on the fly.

To tackle the limitation of the extra processing requirements for performing join operations in the MapReduce framework, the *Map-Reduce-Merge* model [23] have been introduced to enable the processing of multiple datasets. Figure 2.4 illustrates the framework of this model where the map phase transforms an input key/value pair $(k1, v1)$ into a list of intermediate key/value pairs $[(k2, v2)]$. The Reduce function aggregates the list of values $[v2]$ associated with $k2$ and produces a list of values $[v3]$ which is also associated with $k2$. Note that inputs and outputs of both functions belong to the same lineage (α). Another pair of Map and Reduce functions produce the intermediate output $(k3, [v4])$ from another lineage (β). Based on keys $k2$ and $k3$, the merge function combines the two reduced outputs from different lineages into a list of key/value outputs $[(k4, v5)]$. This final output becomes a new lineage (γ). If $\alpha = \beta$ then this merge function does a self-merge which is similar to self-join

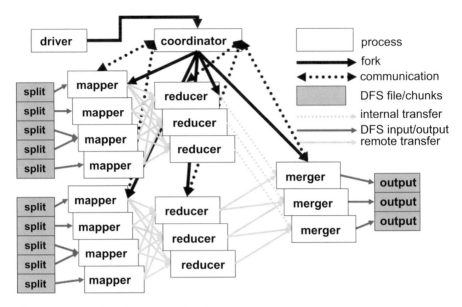

Fig. 2.4 An overview of the Map-Reduce-Merge framework [23]

in relational algebra. The main differences between the processing model of this framework and the original MapReduce are the production of a key/value list from the Reduce function instead of just that of values. This change is introduced because the merge function requires input datasets to be organized (partitioned, then either sorted or hashed) by keys and these keys have to be passed into the function to be merged. In the original framework, the reduced output is final. Hence, users pack whatever is needed in [$v3$] while passing $k2$ for the next stage is not required.

The *Map-Join-Reduce* [26] represents another approach that has been introduced with a filtering-join-aggregation programming model as an extension of the standard MapReduce's filtering-aggregation programming model. In particular, in addition to the standard mapper and reducer operation of the standard MapReduce framework, they introduce a third operation, join (called joiner), to the framework. Hence, to join multiple datasets for aggregation, users specify a set of *join()* functions and the join order between them. Then, the runtime system automatically joins the multiple input datasets according to the join order and invoke *join()* functions to process the joined records. They have also introduced a one-to-many shuffling strategy which shuffles each intermediate key/value pair to many Joiners at one time. Using a tailored partition strategy, they can utilize the one-to-many shuffling scheme to join multiple datasets in one phase instead of a sequence of MapReduce jobs. The runtime system for executing a Map-Join-Reduce job launches two kinds of processes: *MapTask* and *ReduceTask*. Mappers run inside the MapTask process while Joiners and Reducers are invoked inside the ReduceTask process. Therefore, Map-Join-Reduce's process model allows for the pipelining of intermediate results between Joiners and Reducers since Joiners and Reducers are run inside the same ReduceTask process.

2.1.2.2 Supporting Iterative Processing

The basic MapReduce framework does not directly support these iterative data analysis applications. Instead, programmers must implement iterative programs by manually issuing multiple MapReduce jobs and orchestrating their execution using a driver program. In practice, there are two key problems with manually orchestrating an iterative program in MapReduce:

- Even though much of the data may be unchanged from iteration to iteration, the data must be reloaded and reprocessed at each iteration, wasting I/O, network bandwidth, and CPU resources.
- The termination condition may involve the detection of when a fix point has been reached. This condition may itself require an extra MapReduce job on each iteration, again incurring overhead in terms of scheduling extra tasks, reading extra data from disk, and moving data across the network.

The *HaLoop* system [27] is designed to support iterative processing on the MapReduce framework by extending the basic MapReduce framework with two main functionalities:

1. Caching the invariant data in the first iteration and then reusing them in later iterations.
2. Caching the reducer outputs, which makes checking for a fix point more efficient, without an extra MapReduce job.

Figure 2.5 illustrates the architecture of HaLoop as a modified version of the basic MapReduce framework. In principle, HaLoop relies on the same file system and has the same task queue structure as Hadoop but the task scheduler and task tracker modules are modified, and the loop control, caching, and indexing modules are newly introduced to the architecture. The task tracker not only manages task execution but also manages caches and indices on the slave node and redirects each task's cache and index accesses to local file system.

In the MapReduce framework, each map or reduce task contains its portion of the input data and the task runs by performing the Map/Reduce function on its input data records where the life cycle of the task ends when finishing the processing of all the input data records has been completed. The *iMapReduce* framework [28] supports the feature of iterative processing by keeping alive each map and reduce task during the whole iterative process. In particular, when all of the input data of a persistent task are parsed and processed, the task becomes dormant, waiting for the new updated input data. For a map task, it waits for the results from the reduce tasks and is activated to work on the new input records when the required data from the reduce tasks arrive. For the reduce tasks, they wait for the map tasks' output and are

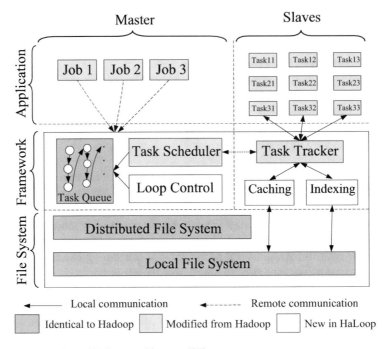

Fig. 2.5 An overview of HaLoop architecture [27]

activated synchronously as in MapReduce. Jobs can terminate their iterative process in one of two ways:

1. *Defining fixed number of iterations*: Iterative algorithm stops after it iterates n times.
2. *Bounding the distance between two consecutive iterations*: Iterative algorithm stops when the distance is less than a threshold.

The iMapReduce runtime system does the termination check after each iteration. To terminate the iterations by a fixed number of iterations, the persistent map/reduce task records its iteration number and terminates itself when the number exceeds a threshold. To bound the distance between the output from two consecutive iterations, the reduce tasks can save the output from two consecutive iterations and compute the distance. If the termination condition is satisfied, the master will notify all the map and reduce tasks to terminate their execution.

Other projects have been implemented for supporting iterative processing on the MapReduce framework. For example, *Twister*[4] is a MapReduce runtime with an extended programming model that supports iterative MapReduce computations

---
[4]http://www.iterativemapreduce.org/.

efficiently [29]. It uses a publish/subscribe messaging infrastructure for communication and data transfers and supports long running map/reduce tasks. In particular, it provides programming extensions to MapReduce with broadcast and scatter type data transfers. Microsoft has also developed a project that provides an iterative MapReduce runtime for Windows Azure called *Daytona*.[5]

2.1.2.3 Data and Process Sharing

With the emergence of cloud computing, the use of an analytical query processing infrastructure (e.g., Amazon EC2) can be directly mapped to *monetary* value. Taking into account that different MapReduce jobs can perform similar work, there could be many opportunities for sharing the execution of their work. Thus, this sharing can reduce the overall amount of work which consequently leads to the reduction of the monetary charges incurred while utilizing the resources of the processing infrastructure. The *MRShare* system [30] have been presented as a sharing framework which is tailored to transform a batch of queries into a new batch that will be executed more efficiently by merging jobs into groups and evaluating each group as a single query. Based on a defined cost model, they described an optimization problem that aims to derive the optimal grouping of queries in order to avoid performing redundant work and thus resulting in significant savings on both processing time and money. While the *MRShare* system focuses on sharing the processing between queries that are executed concurrently, the *ReStore* system [31, 32] has been introduced so that it can enable the queries that are submitted at different times to share the intermediate results of previously executed jobs and reusing them for future submitted jobs to the system. In particular, each MapReduce job produces output that is stored in the distributed file system used by the MapReduce system (e.g., HDFS). These intermediate results are kept (for a defined period) and managed so that they can be used as input by subsequent jobs. ReStore can make use of whole jobs or sub-jobs reuse opportunities.

2.1.2.4 Support of Data Indices and Column Storage

One of the main limitations of the original implementation of the MapReduce framework is that it is designed in a way that the jobs can only scan the input data in a sequential-oriented fashion. Hence, the query processing performance of the MapReduce framework is unable to match the performance of a well-configured parallel DBMS [19]. In order to tackle this challenge, the *Hadoop++* system [33] introduced the following main changes:

- *Trojan Index*: The original Hadoop implementation does not provide index access due to the lack of a priori knowledge of the schema and the MapReduce jobs being executed. Hence, the Hadoop++ system is based on the assumption that if we know the schema and the anticipated MapReduce jobs, then we can create appropriate indices for the Hadoop tasks. In particular, Trojan index is

---

[5]http://research.microsoft.com/en-us/projects/daytona/.

an approach to integrate indexing capability into Hadoop in a non-invasive way. These indices are created during the data loading time and thus have no penalty at query time. Each Trojan index provides an optional index access path which can be used for selective MapReduce jobs.

- *Trojan Join*: Similar to the idea of the Trojan index, the Hadoop++ system assumes that if we know the schema and the expected workload, then we can co-partition the input data during the loading time. In particular, given any two input relations, they apply the same partitioning function on the join attributes of both the relations at data loading time and place the co-group pairs, having the same join key from the two relations, on the same split and hence on the same node. As a result, join operations can be then processed locally within each node at query time. Implementing the Trojan joins does not require any changes to be made to the existing implementation of the Hadoop framework. The only changes are made on the internal management of the data splitting process. In addition, Trojan indices can be freely combined with Trojan joins.

The design and implementation of a column-oriented and binary back-end storage format for Hadoop has been presented in [34]. In general, a straightforward way to implement a column-oriented storage format for Hadoop is to store each column of the input dataset in a separate file. However, this raises two main challenges:

- It requires generating roughly equal sized splits so that a job can be effectively parallelized over the cluster.
- It needs to ensure that the corresponding values from different columns in the dataset are co-located on the same node running the map task.

The first challenge can be tackled by horizontally partitioning the dataset and storing each partition in a separate subdirectory. The second challenge is harder to tackle because of the default 3-way block-level replication strategy of HDFS that provides fault tolerance on commodity servers but does not provide any co-location guarantees. Floratou et al. [34] tackle this challenge by implementing a modified HDFS block placement policy which guarantees that the files corresponding to the different columns of a split are always co-located across replicas. Hence, when reading a dataset, the column input format can actually assign one or more split-directories to a single split and the column files of a split-directory are scanned sequentially where the records are reassembled using values from corresponding positions in the files. A lazy record construction technique is used to mitigate the deserialization overhead in Hadoop, as well as eliminate unnecessary disk I/O. The basic idea behind lazy record construction is to deserialize only those columns of a record that are actually accessed in a Map function. One advantage of this approach is that adding a column to a dataset is not an expensive operation. This can be done by simply placing an additional file for the new column in each of the split-directories. On the other hand, a potential disadvantage of this approach is that the available parallelism may be limited for smaller datasets. Maximum parallelism is achieved for a MapReduce job when the number of splits is at least equal to the number of map tasks.

The *Llama* system [35] has introduced another approach of providing column storage support for the MapReduce framework. In this approach, each imported table is transformed into column groups where each group contains a set of files representing one or more columns. Llama introduced a column-wise format for Hadoop, called *CFile*, where each file can contain multiple data blocks and each block of the file contains a fixed number of records. However, the size of each logical block may vary since records can be variable-sized. Each file includes a block index, which is stored after all data blocks, stores the offset of each block, and is used to locate a specific block. In order to achieve storage efficiency, Llama uses block-level compression by using any of the well-known compression schemes. In order to improve the query processing and the performance of join operations, Llama columns are formed into correlation groups to provide the basis for the vertical partitioning of tables. In particular, it creates multiple vertical groups where each group is defined by a collection of columns, one of them is specified as the sorting column. Initially, when a new table is imported into the system, a basic vertical group is created which contains all the columns of the table and sorted by the table's primary key by default. In addition, based on statistics of query patterns, some auxiliary groups are dynamically created or discarded to improve the query performance. The *Clydesdale* system [36,37], a system which has been implemented for targeting workloads where the data fits a star schema, uses *CFile* for storing its fact tables. It also relies on tailored join plans and block iteration mechanism [38] for optimizing the execution of its target workloads.

RCFile [39] (Record Columnar File) is another data placement structure that provides column-wise storage for Hadoop Distributed File System (HDFS). In RCFile, each table is firstly stored by horizontally partitioning it into multiple row groups where each row group is then vertically partitioned so that each column is stored independently. In particular, each table can have multiple HDFS blocks where each block organizes records with the basic unit of a row group. Depending on the row group size and the HDFS block size, an HDFS block can have only one or multiple row groups. In particular, a row group contains the following three sections:

1. The *sync marker* which is placed in the beginning of the row group and mainly used to separate two continuous row groups in an HDFS block.
2. A metadata header which stores the information items on how many records are in this row group, how many bytes are in each column, and how many bytes are in each field in a column.
3. The table data section which is actually a column-store where all the fields in the same column are stored continuously together.

RCFile utilizes a column-wise data compression within each row group and provides a lazy decompression technique to avoid unnecessary column decompression during query execution. In particular, the metadata header section is compressed using the *RLE* (Run Length Encoding) algorithm. The table data section is not compressed as a whole unit. However, each column is independently compressed with the *Gzip* compression algorithm. When processing a row group, RCFile does not need to fully read the whole content of the row group into memory. However, it only reads the metadata header and the needed columns in the row group for a given query and thus it can skip unnecessary columns and gain the I/O advantages of a column-store. The metadata header is always decompressed and held in memory until RCFile processes the next row group. However, RCFile does not decompress all the loaded columns and uses a lazy decompression technique where a column will not be decompressed in memory until RCFile has determined that the data in the column will be really useful for query execution.

The notion of *Trojan Data Layout* has been coined in [40] which exploits the existing data block replication in HDFS to create different Trojan Layouts on a per-replica basis. This means that rather than keeping all data block replicas in the same layout, it uses *different* Trojan Layouts for each replica which is optimized for a different subclass of queries. As a result, every incoming query can be scheduled to the most suitable data block replica. In particular, Trojan Layouts change the internal organization of a data block and not among data blocks. They co-locate attributes together according to query workloads by applying a column grouping algorithm which uses an interestingness measure that denotes how well a set of attributes speeds up most or all queries in a workload. The column groups are then packed in order to maximize the total interestingness of data blocks. At query time, an incoming MapReduce job is transparently adapted to query the data block replica that minimizes the data access time. The map tasks are then routed of the MapReduce job to the datanodes storing such data block replicas.

2.1.2.5 Effective Data Placement

In the basic implementation of the Hadoop project, the objective of the data placement policy is to achieve good load balancing by distributing the data evenly across the data servers, independently of the intended use of the data. This simple data placement policy works well with most Hadoop applications that access just a *single* file. However, there are some other applications that process data from *multiple* files which can get a significant boost in performance with customized strategies. In these applications, the absence of data co-location increases the data shuffling costs, increases the network overhead, and reduces the effectiveness of data partitioning. *CoHadoop* [41] is a lightweight extension to Hadoop which is designed to enable co-locating related files at the file system level while at the same time retaining the good load balancing and fault tolerance properties. It introduces a new file property to identify related data files and modify the data placement policy of Hadoop to co-locate copies of those related files in the same server. These changes are designed in a way to retain the benefits of Hadoop, including load balancing and fault tolerance. In principle, CoHadoop provides a generic mechanism that allows applications to control data placement at the file system level. In particular, a new file-level property called a *locator* is introduced and the Hadoop's data placement policy is modified so that it makes use of this locator property. Each locator is

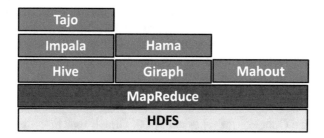

Fig. 2.6 Hadoop's ecosystem

represented by a unique value (ID) where each file in HDFS is assigned to at most one locator and many files can be assigned to the same locator. Files with the same locator are placed on the same set of datanodes, whereas files with no locator are placed via Hadoop's default strategy. It should be noted that this co-location process involves all data blocks, including replicas.

2.1.3   Hadoop's Ecosystem

Over the years, the Hadoop's ecosystem has been extended with various components (Fig. 2.6). For example, the over simplicity of the MapReduce programming model and its reliance on a rigid one-input and two-stage dataflow lead to situations in which inelegant workarounds are required when performing tasks that have a different workflow (e.g., joins or n stages). The *Hive* project [42] has been introduced to support SQL-on-Hadoop with familiar relational database concepts such as tables, columns, and partitions. It supports queries that are expressed in an SQL-like declarative language, Hive Query Language (HiveQL[6]), which represents a subset of SQL92, and therefore can be easily understood by anyone who is familiar with SQL. These queries automatically compile into Hadoop jobs. *Impala*[7] is another open-source project, built by Cloudera, to provide a massively parallel processing SQL query engine that runs natively in Apache Hadoop. It utilizes the standard components of Hadoop's infrastructure (e.g., HDFS, HBase, YARN) and is able to read the majority of the widely used file formats (e.g., Parquet, Avro). Therefore, by using Impala, the user can query data which is stored in Hadoop Distributed File System (HDFS). The IBM big data processing platform, InfoSphere BigInsights, which is built on the Apache Hadoop framework has provided *Big SQL* engine as its SQL interface. In particular, it provides SQL access to data that is stored in InfoSphere BigInsights and uses the Hadoop framework for complex datasets and

---

[6]https://cwiki.apache.org/confluence/display/Hive/LanguageManual.
[7]http://impala.io/.

direct access for smaller queries. *Apache Tajo*[8] is another distributed data warehouse system for Apache Hadoop that is designed for low-latency and scalable ad-hoc queries ETL processes. Tajo can analyze data which is stored on HDFS, Amazon S3, OpenStack Swift,[9] and local file systems. It provides an extensible query re-write system that lets users and external programs query data through SQL.

Apache Giraph is another component which has been introduced as an open-source project that supports large-scale graph processing and clones the implementation of the Google's *Pregel* system [43]. Giraph runs graph processing jobs as map-only jobs on Hadoop and uses HDFS for data input and output. *Apache Hama*[10] is another BSP-based implementation project which is designed to run on top of the Hadoop infrastructure, like Giraph. However, it focuses on general BSP computations and not only on graph processing. For example, it includes algorithms for matrix inversion and linear algebra. Machine learning algorithms represent another type of applications which are iterative in nature. *Apache Mahout*[11] project has been designed for building scalable machine learning libraries on top of the Hadoop framework.

2.2   Spark

The Hadoop framework has pioneered the domain of general purpose data processing systems. However, one of the main performance limitations of the Hadoop framework is that it materializes the intermediate results of each Map or Reduce step on HDFS before starting the next one (Fig. 2.7). Thus, several systems have been developed to tackle the performance problem of the Hadoop framework. The Spark project has been introduced as a general-purpose big data processing engine which can be used for many types of data processing scenarios [44]. In principle, Spark was initially designed for providing efficient performance for interactive queries and iterative algorithms, as these were two major use cases which were not well-supported by the MapReduce framework. Spark, written in Scala[12] [45], was originally developed in the AMPLab at UC Berkeley[13] and open-sourced in 2010 as one of the new generation dataflow engines that was developed to overcome the limitations of the MapReduce framework.

In general, one of the main limitations of the Hadoop framework is that it requires that the entire output of each map and reduce task to be materialized into a local file on the Hadoop Distributed File System (HDFS) before it can be consumed by the next stage. This materialization step allows for the implementation of a simple and elegant checkpoint/restart fault tolerant mechanism; however, it dramatically

---

[8]http://tajo.apache.org/.

[9]http://docs.openstack.org/developer/swift/.

[10]https://hama.apache.org/.

[11]http://mahout.apache.org/.

[12]http://www.scala-lang.org/.

[13]https://amplab.cs.berkeley.edu/.

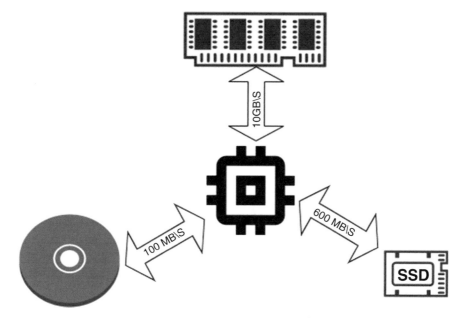

Fig. 2.7 Differences on data transfer speed

harms the system performance. Spark takes the concepts of Hadoop to the next level by loading the data in distributed memory and relying on less expensive shuffles during the data processing (Fig. 2.8). In particular, the fundamental programming abstraction of Spark is called *Resilient Distributed Datasets* (RDD) [44] which represents a logical collection of data partitioned across machines that are created by referencing datasets in external storage systems, or by applying various and rich coarse-grained transformations (e.g., *map, filter, reduce, join*) on existing RDDs rather than the two simple programming abstractions, *map* and *reduce*, of the Hadoop framework. For instance, the *map* transformation of Spark applies a transformation function to each element in the input RDD and returns a new RDD with the elements of the transformation output as a result, the *filter* transformation applies a filtration predicate on the elements of an RDD and returns a new RDD with only the elements which satisfy the predicate conditions, while the *union* transformation returns all elements of two input RDDs in a new RDD. In principle, having RDD as in-memory data structure gives the power to Spark's functional programming paradigm by allowing user programs to load data into a cluster's memory and query it repeatedly. In addition, users can explicitly cache an RDD in memory across machines and reuse it in multiple MapReduce-like parallel operations. In particular, RDDs can be manipulated through operations like map, filter, and reduce, which take functions in the programming language and ship them to nodes on the cluster. This simplifies programming complexity because the way applications manipulate RDDs is similar to that of manipulating local collections of

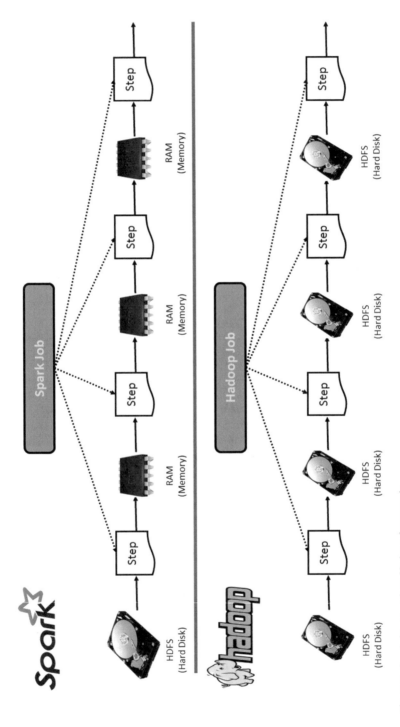

Fig. 2.8 Spark framework vs Hadoop framework

data. During the execution phase, Spark follows a lazy mechanism for the evaluation of program operations where RDDs are not always immediately materialized. Instead, data in RDDs is not processed and is not materialized in memory until an *action* function (e.g., *first*, *count*, *countByValye*, *reduce*) is performed over it then Spark engine launches a computation to materialize the new RDD. For example, the *first* action returns the first element in an RDD, the *count* action returns the number of elements in an RDD, while the *reduce* action combines the elements on an RDD together according to an aggregate function.

RDDs achieve fault tolerance through a notion of lineage so that a resilient distributed dataset (RDD) can be rebuilt if a partition is lost [44]. In other words, instead of relying on schemes for persisting or checkpointing intermediate results, Spark remembers the sequence of operations which led to a certain dataset. In particular, if a partition of an RDD is lost, the RDD has enough information about how it was derived from other RDDs to be able to rebuild just that partition. Spark is built on top of *Mesos* [46], a cluster operating system that lets multiple parallel frameworks share a cluster in a fine-grained manner and provides an API for applications to launch tasks on a cluster. It provides isolation and efficient resource sharing across frameworks running on the same cluster while giving each framework freedom to implement its own programming model and fully control the execution of its jobs. Mesos uses two main abstractions: *tasks* and *slots*. A task represents a unit of work. A slot represents a computing resource in which a framework may run a task, such as a core and some associated memory on a multicore machine. It employs the two-level scheduling mechanism. At the first level, Mesos allocates slots between frameworks using fair sharing. At the second level, each framework is responsible for dividing its work into tasks, selecting which tasks to run in each slot. This lets frameworks perform application-specific optimizations. For example, Spark's scheduler tries to send each task to one of its preferred locations using a technique called *delay scheduling* [47].

To use Spark, developers need to write a driver program that implements the high-level control flow of their application and launches various operations in parallel. Spark provides two main abstractions for parallel programming: resilient distributed datasets and parallel operations on these datasets (invoked by passing a function to apply on a dataset). In particular, each RDD is represented by a Scala object which can be constructed in different ways:

- From a file in a shared file system (e.g., HDFS).
- By parallelizing a Scala collection (e.g., an array) in the driver program which means dividing it into a number of slices that will be sent to multiple nodes.
- By transforming an existing RDD. A dataset with elements of type A can be transformed into a dataset with elements of type B using an operation called $flatMap$.
- By changing the persistence of an existing RDD. A user can alter the persistence of an RDD through two actions:
  - The cache action leaves the dataset lazy but hints that it should be kept in memory after the first time it is computed because it will be reused.

– The save action evaluates the dataset and writes it to a distributed file system such as HDFS. The saved version is used in future operations on it.

Different parallel operations can be performed on RDDs:

- The *reduce* operation which combines dataset elements using an associative function to produce a result at the driver program.
- The *collect* operation which sends all elements of the dataset to the driver program.
- The *foreach* which passes each element through a user-provided function.

Spark does not currently support a grouped reduce operation as in MapReduce. The results of reduce operations are only collected at the driver process. In principle, the RDD abstraction enables developers to materialize any point in a processing pipeline into memory across the cluster, meaning that future steps that want to deal with the same dataset need not recompute it or reload it from disk. In addition, developers can explicitly cache an RDD in memory across machines and reuse it in multiple parallel operations. Thus, Spark is well suited for highly iterative algorithms that require multiple passes over a dataset, as well as reactive applications that quickly respond to user queries by scanning large in-memory datasets. This has been considered one of the main bottleneck scenarios for the Hadoop framework.

In practice, developers can create RDDs in various ways. For example, RDD can be created by loading data from a file or a set of files where each line in the file (s) is represented as a separate record in the RDD. The following code snippet shows an example of loading file (s) into RDD using the *SparkContext*:

```
// single file
sc.textFile("mydata.txt")
//comma-separated list of files
sc.textFile("mydatafile1.txt, mydatafile2.txt ")
// a wildcare list of file
sc.textFile("mydirectory/*.txt")
```

Spark can interface with a wide variety of distributed storage implementations, including Hadoop Distributed File System (HDFS) [48], Cassandra,[14] and Amazon S3.[15] RDD can be also created by distributing a collection of objects (e.g., a list or set) which are loaded in memory or by applying coarse-grained transformations (e.g., map, filter, reduce, join) on existing RDDs. Spark depends heavily on the concepts of functional programming. In Spark, functions represent the fundamental unit of programming where functions can only have input and output but with no state or side effects. In principle, Spark offers two types of operations over RDDs:

---

[14]http://cassandra.apache.org/.

[15]http://aws.amazon.com/s3/.

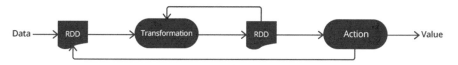

Fig. 2.9 Flow of RDD operations in Spark

Table 2.1 Sample Spark's transformations

Transformation	Description
map	Apply a transformation function to each element in the input RDD and returns a new RDD with the elements of the transformation output as a result
filter	Apply a filtration predicate on the elements of an RDD and returns a new RDD with only the elements which satisfy the predicate conditions
distinct	Remove the duplicate elements of an RDD
union	Return all elements of two RDDs
Cartesian	Return the Cartesian product of the elements of two RDDs
intersection	Return the elements which are contained in two RDDs
subtract	Return the elements which are not contained in another RDD

Table 2.2 Sample Spark's actions

Action	Description
take	Return number of elements from an RDD
takeOrdered	Return number of elements from an RDD based on defined order
top	Return the top number of elements from an RDD
count	Return the number of elements in an RDD
countByValye	Return the number of times each element occurs in an RDD
reduce	Combine the elements on an RDD together according to an aggregate function
foreach	Apply a function for each element in an RDD

transformations and *actions*. Transformations are used to construct a new RDD from an existing one. For example, one common transformation is filtering data that matches a predicate. On the other hand, Actions is used to compute a result based on an existing RDD and return the results either to the driver program or save it to an external storage system (e.g., HDFS). Figure 2.9 illustrates the flow of RDD operations in Spark. Table 2.1 provides an overview of some Spark's transformation while Table 2.2 provides an overview of some Spark's actions

Other than Spark Core API, there are additional libraries that are part of the Spark ecosystem and provide additional capabilities in the Big Data processing area (Fig. 2.10). In particular, Spark provides various packages with higher-level libraries including support for SQL queries [49], streaming data,[16] machine learning [50], statistical programming, and graph processing [51]. These libraries increase devel-

---

[16]https://spark.apache.org/streaming/.

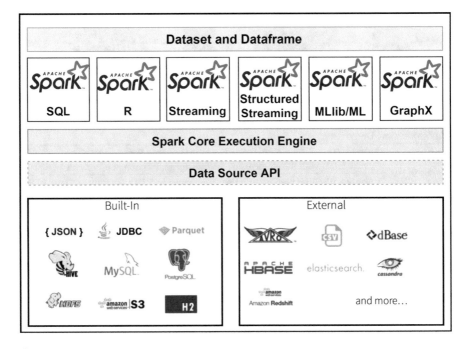

Fig. 2.10 Spark's ecosystem

oper productivity and can be seamlessly combined to create complex workflows. In addition, Spark provides APIs for various programming languages including Scala, Java, Python, and R. In the 2014 annual Daytona Gray Sort Challenge,[17] which benchmarks the speed of data analysis systems, Spark has strongly outperformed Hadoop and was able to sort through 100 terabytes of records within 23 min, while Hadoop took over three times as long to execute the same task, about 72 min. Currently, Spark has over 500 contributors from more than 200 organizations making it the most active project in the Apache Software Foundation and among Big Data open-source projects. Popular distributors of the Hadoop ecosystem (e.g., Cloudera, Hortonworks, and MapR) are currently including Spark in their releases.

2.3   Flink

Apache[18] Flink[19] is another distributed in-memory data processing framework which represents a flexible alternative for the MapReduce framework that supports both of batch and real-time processing. Instead of the map and reduce abstractions of

---

[17]http://sortbenchmark.org/.

[18]https://flink.apache.org/.

[19]Flink is a German word that means "quick" or "nimble".

Fig. 2.11 PACT
programming model [53]

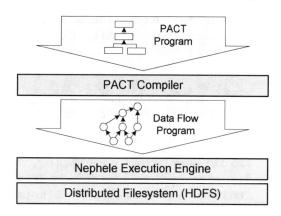

the MapReduce framework, Flink uses a directed graph approach that leverages in-memory storage for improving the performance of the runtime execution. Flink can run as a completely independent framework but can run on top of HDFS and YARN as well. Flink can also directly allocate hardware resources from infrastructure-as-a-service clouds, such as Amazon EC2. Currently, Flink provides programming APIs for Scala and Java, as well as Spargel, a specialized API that implements a Pregel bulk synchronous parallel programming model. Recently, Flink has become a top level project at the Apache open-source software foundation.

In principle, Flink has originated from the *Stratosphere* research project[20] that was started at the Technical University of Berlin in 2009 before joining the Apache's incubator in 2014 [52]. Stratosphere uses a richer set of primitives than MapReduce, including primitives that allow the easy specification, automatic optimization, and efficient execution of joins. It treats user-defined functions (UDFs) as first-class citizens and relies on a query optimizer that automatically parallelizes and optimizes big data processing jobs. Stratosphere offers both pipeline (inter-operator) and data (intra-operator) parallelism. In particular, Stratosphere relies on the *Parallelization Contracts* (PACTs) programming model [53, 54] which represents a generalization of map/reduce as it is based on a key/value data model and the concept of *Parallelization Contracts* (PACTs). A PACT consists of exactly one second-order function which is called *Input Contract* and an optional *Output Contract*. An Input Contract takes a first-order function with task-specific user code and one or more datasets as input parameters (Fig. 2.11). The Input Contract invokes its associated first-order function with independent subsets of its input data in a data-parallel fashion. In this context, the *only* two functions of the Hadoop framework, *map* and *reduce*, are just examples of the Input Contracts. Other example of Input Contracts includes:

- The *Cross* contract which operates on multiple inputs and builds a distributed Cartesian product over its input sets.

---

[20]http://stratosphere.eu/.

- The *CoGroup* contract partitions each of its multiple inputs along the key. Independent subsets are built by combining equal keys of all inputs.
- The *Match* contract operates on multiple inputs. It matches key/value pairs from all input datasets with the same key (equivalent to the inner join operation).

An Output Contract is an optional component of a PACT and gives guarantees about the data that is generated by the assigned user function. The set of Output Contracts includes:

- The *Same-Key* contract where each key/value pair that is generated by the function has the same key as the key/value pair(s) from which it was generated. This means the function will preserve any partitioning and order property on the keys.
- The *Super-Key* where each key/value pair that is generated by the function has a super-key of the key/value pair(s) from which it was generated. This means the function will preserve a partitioning and partial order on the keys.
- The *Unique-Key* where each key/value pair that is produced has a unique key. The key must be unique across all parallel instances. Any produced data is therefore partitioned and grouped by the key.
- The *Partitioned-by-Key* where key/value pairs are partitioned by key. This contract has similar implications as the Super-Key contract, specifically that a partitioning by the keys is given, but there is no order inside the partitions.

Stratosphere uses an execution engine, *Nephele*, that supports external memory query processing algorithms and natively supports arbitrarily long programs shaped as directed acyclic graphs [53, 54]. In Stratosphere, a PACT program is submitted to the PACT Compiler which translates the program into a dataflow execution plan which is then handed to the Nephele system for parallel execution [53, 54]. In particular, the incoming jobs of Nephele are represented as dataflow graphs where vertices represent subtasks and edges represent communication channels between these subtasks. Each subtask is a sequential program which reads data from its input channels and writes to its output channels. Prior to an execution, Nephele generates the parallel dataflow graph by spanning the received DAG. Hence, vertices are multiplied to the desired degree of parallelism. Connection patterns that are attached to channels define how the multiplied vertices are rewired after spanning. During execution, the Nephele system takes care of resource scheduling, task distribution, communication as well as synchronization issues. Moreover, Nephele's fault tolerance mechanisms help to mitigate the impact of hardware outages. Nephele also offers the ability to annotate the input jobs with a rich set of parameters that could influence the physical execution. For example, it is possible to set the desired degree of data parallelism for each subtask, assign particular sets of subtasks to particular sets of compute nodes, or explicitly specify the type of communication channels between subtasks. Nephele also supports three different types of communication channels: network, in-memory, and file channels. The network and in-memory channels allow the PACT compiler to construct low-

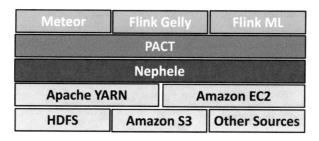

Fig. 2.12 Flink software stack

latency execution pipelines in which one task can immediately consume the output of another. The file channels collect the entire output of a task in a temporary file before passing its content on to the next task. Therefore, file channels can be considered as checkpoints, which help to recover from execution failures.

Due to the declarative character of the PACT programming model, the PACT compiler can apply different optimization mechanisms and select from several execution plans with varying costs for a single PACT program. For example, the *Match* contract can be satisfied using either a repartition strategy which partitions all inputs by keys or a broadcast strategy that fully replicates one input to every partition of the other input. Choosing the right strategy can dramatically reduce network traffic and execution time. Therefore, the PACT compiler applies standard SQL optimization techniques [55] where it exploits information provided by the Output Contracts and applies different cost-based optimization techniques. In particular, the optimizer generates a set of candidate execution plans in a bottom-up fashion (starting from the data sources) where the more expensive plans are pruned using a set of *interesting properties* for the operators. These properties are also used to spare plans from pruning that come with an additional property that may amortize their cost overhead later.

Figure 2.12 illustrates the layered architecture of Flink that provides various packages and programming interfaces for its end users. For example, Stratosphere provides *Meteor*, an operator-oriented query language that uses a JSON-like data model to support the analysis of unstructured and semi-structured data [56]. In particular, Meteor shares similar objectives as higher-level languages of other big data stacks (e.g., Pig and Hive for Hadoop and SparkSQL for Spark). When a Meteor script get submitted to Stratosphere, it is first translated into an operator plan which goes through an optimization process before being compiled into a PACT program.

The Flink systems is equipped with *Flink Streaming*[21] API as an extension of the core Flink API for high-throughput and low-latency data stream processing. The system can connect to and process data streams from various data sources (e.g., Flume, ZeroMQ) where data streams can be transformed and modified using high-

---

[21] http://ci.apache.org/projects/flink/flink-docs-release-0.7/streaming_guide.html.

level functions similar to the ones provided by the batch processing API. In addition, the Flink open-source community has developed libraries for machine learning (FlinkML[22]) and graph processing (Flink Gelly[23]). In 2019, Alibaba acquired Apache Flink project[24] and maintained its led extension as a closed product with the name *ververica*.[25]

2.4   Hyracks/ASTERIX

Hyracks has been presented as a partitioned-parallel data flow execution platform that runs on shared-nothing clusters of computers [57]. Large collections of data items are stored as local partitions distributed across the nodes of the cluster. A Hyracks job is submitted by a client and processes one or more collections of data to produce one or more output collections (partitions). Hyracks provides a programming model and an accompanying infrastructure to efficiently divide computations on large data collections (spanning multiple machines) into computations that work on each partition of the data separately. Every Hyracks cluster is managed by a *Cluster Controller* process. The Cluster Controller accepts job execution requests from clients, plans their evaluation strategies, and then schedules the jobs' tasks to run on selected machines in the cluster. In addition, it is responsible for monitoring the state of the cluster to keep track of the resource loads at the various worker machines. The Cluster Controller is also responsible for re-planning and re-executing some or all of the tasks of a job in the event of a failure. On the task execution side, each worker machine that participates in a Hyracks cluster runs a *Node Controller* process. The Node Controller accepts task execution requests from the Cluster Controller and also reports on its health via a heartbeat mechanism.

In Hyracks, data flows between operators over connectors in the form of records that can have an arbitrary number of fields. Hyracks provides support for expressing data-type-specific operations such as comparisons and hash functions. The way Hyracks uses a record as the carrier of data is a generalization of the (key, value) concept of MapReduce. Hyracks strongly promotes the construction of reusable operators and connectors that end users can use to build their jobs. The basic set of Hyracks operators includes:

- The *File Readers/Writers* operators are used to read and write files in various formats from/to local file systems and the HDFS.
- The *Mappers* are used to evaluate a user-defined function on each item in the input.

- The *Sorters* are used to sort input records using user-provided comparator functions.
- The *Joiners* are binary-input operators that perform equi-joins.
- The *Aggregators* are used to perform aggregation using a user-defined aggregation function.

The Hyracks connectors are used to distribute data produced by a set of sender operators to a set of receiver operators. The basic set of Hyracks connectors includes:

- The *M:N Hash-Partitioner* hashes every tuple produced by senders to generate the receiver number to which the tuple is sent. Tuples produced by the same sender keep their initial order on the receiver side.
- The *M:N Hash-Partitioning Merger* takes as input sorted streams of data and hashes each tuple to find the receiver. On the receiver side, it merges streams coming from different senders based on a given comparator and thus producing ordered partitions.
- The *M:N Range-Partitioner* partitions data using a specified field in the input and a range-vector.
- The *M:N Replicator* copies the data produced by every sender to every receiver operator.
- The *1:1 Connector* connects exactly one sender to one receiver operator.

In principle, Hyracks has been designed with the goal of being a runtime platform where users can create their jobs and also to serve as an efficient target for the compilers of higher-level programming languages such as Pig, Hive, or Jaql. The *ASTERIX* project[26] uses this feature with the aim of building a scalable information management system that supports the storage, querying, and analysis of large collections of semi-structured nested data objects [58–60]. The ASTERIX data storage and query processing are based on its own semi-structured model called the *ASTERIX Data Model* (ADM). Each individual ADM data instance is typed and self-describing. All data instances live in *datasets* (the ASTERIX analogy to tables) and datasets can be indexed, partitioned and possibly replicated to achieve the scalability and availability goals. External datasets which reside in files that are not under ASTERIX control are also supported. An instance of the ASTERIX data model can either be a primitive type (e.g., integer, string, time) or a derived type, which may include:

- *Enum*: An enumeration type, whose domain is defined by listing the sequence of possible values.
- *Record*: A set of fields where each field is described by its name and type. A record can be either an open record where it contains fields that are not part of

---

[26]https://wiki.apache.org/incubator/AsterixDBProposal.

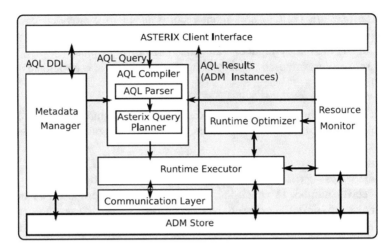

Fig. 2.13 The ASTERIX system architecture [58]

the type definition or a closed record which can only contain fields that are part of the type definition.

- *Ordered list*: A sequence of values for which the order is determined by creation or insertion.
- *Unordered list*: An unordered sequence of values which is similar to bags in SQL.
- *Union*: describes A choice between a finite set of types.

Figure 2.13 presents an overview of the ASTERIX system architecture. AQL (ASTERIX Query Language) requests are compiled into jobs for an ASTERIX execution layer, Hyracks. The top most layer of the ASTERIX architecture is concerned with the data details of AQL and ADM, turning AQL requests into Hyracks jobs while Hyracks determines and oversees the utilization of parallelism based on information and constraints associated with the resulting jobs' operators as well as on the runtime state of the cluster.

A dataset is a target for AQL queries and updates and is also the attachment point for indexes. A collection of datasets related to an application are grouped into a namespace called a *dataverse* which is analogous to a database in the relational world. In particular, data is accessed and manipulated through the use of the *ASTERIX Query Language* (AQL) which is designed to cleanly match and handle the data structuring constructs of ADM. It borrows from *XQuery* and *Jaql* their programmer-friendly declarative syntax that describes bulk operations such as iteration, filtering, and sorting. Therefore, AQL is comparable to those languages in terms of expressive power. The major difference with respect to XQuery is AQL's focus on data-centric use cases at the expense of built-in support for mixed content for document-centric use cases. In ASTERIX, there is no notion of document order or node identity for data instances. Differences between AQL and Jaql stem from the usage of the languages. While ASTERIX data is stored in and managed by the

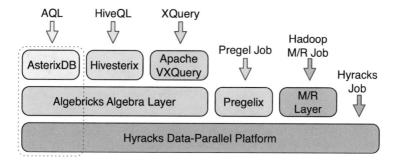

Fig. 2.14 The ASTERIX software stack [60]

Table 2.3 Feature summary of Hadoop, Spark and Flink

	Hadoop	Spark	Flink	ASTERIX
Year of origin	2005	2009	2009	2009
Place of origin	MapReduce (Google) Hadoop (Yahoo)	UC Berkely	TU Berlin	University of California, Riverside
Programming model	Map and Reduce function over key/value pairs	RDD	PACT	Declarative XQuery-based
Data storage	HDFS	HDFS, Cassandra and others	HDFS, S3 and others	HDFS, AsterixDB
SQL support	Hive, Impala, Tajo	Spark SQL	NA	Hivesterix
Graph support	NA	GraphX	Gelly	Pregelix
Real-time streaming support	NA	Spark Streaming	Flink Streaming	NA
Machine learning support	Mahout	Spark MLlib and SparkR	NA	NA

ASTERIX system, Jaql runs against data that are stored externally in Hadoop files or in the local file system. Figure 2.14 illustrates the software stack of the ASTERIX system with various interfaces for SQL (Hivesterix), XQuery (Apache VXQuery[27]), and graph (Pregelix [61]).

Table 2.3 summarizes the features of the various general-purpose data processing frameworks Hadoop, Spark, Flink, and ASTERIX. More details about the SQL interfaces for the various systems will be presented in Chap. 3, the graph interfaces will be covered in Chap. 4, Chap. 5 will describe the stream interfaces while Chap. 6 will describe the machine learning interfaces.

---

[27]https://vxquery.apache.org/.

Large-Scale Processing Systems of Structured Data

3

3.1 Why SQL-On-Hadoop?

For programmers, a key appealing feature in the MapReduce framework is that there are only two main high-level declarative primitives (*map* and *reduce*) that can be written in any programming language of choice and without worrying about the details of their parallel execution. On the other hand, the MapReduce programming model has its own limitations such as:

- Its one-input data format (key/value pairs) and two-stage dataflow are extremely rigid. As we have previously discussed, to perform tasks that have a different dataflow (e.g., joins or n stages) would require the need to devise inelegant workarounds.
- Custom code has to be written for even the most common operations (e.g., projection and filtering) which leads to the fact that the code is usually difficult to reuse and maintain unless the users build and maintain their own libraries with the common functions they use for processing their data.

Moreover, many programmers could be unfamiliar with the MapReduce framework and they would prefer to use SQL (in which they are more proficient) as a high-level declarative language to express their task while leaving all of the execution optimization details to the back-end engine. In addition, it is beyond doubt that high-level language abstractions enable the underlying system to perform automatic optimization. In the following subsection we discuss research efforts that have been proposed to tackle these problems and add SQL-like interfaces on top of the MapReduce framework.

S. Sakr, *Big Data 2.0 Processing Systems*, https://doi.org/10.1007/978-3-030-44187-6_3

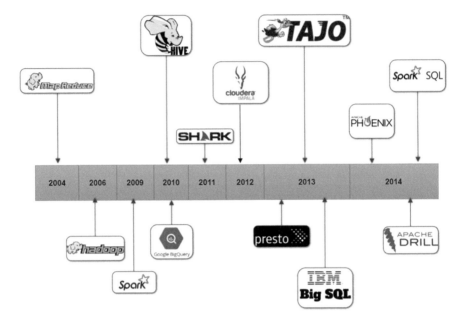

Fig. 3.1 Timeline of SQL-on-Hadoop systems

One of the main limitations of the programming model of the Hadoop framework is that it relies on a simple one-input and two-stage dataflow, and, thus, performing tasks that have a different dataflow (e.g., joins or n stages) would require implementing inelegant workarounds. On the other hand, various studies (e.g., [62]) showed that Hadoop does not represent the right solution for dealing with workloads that involve interactive queries with a target response time of a few milliseconds or seconds. Particularly, traditional parallel database systems have shown to be significantly outperforming in the domain of processing large-scale structured data.

To address the above challenges, various solutions have been introduced to provide the SQL flavor on top of the big data processing platforms so that they can provide competing and scalable performance on processing large-scale structured data (Fig. 3.1).

3.2   Hive

The *Hive* project[1] is an open-source data warehousing solution which has been built by the Facebook Data Infrastructure Team on top of the Hadoop environment [63]. The main goal of this project is to bring the familiar relational database concepts

---

[1] http://hadoop.apache.org/hive/.

```
FROM (
    MAP doctext USING 'python wc_mapper.py' AS (word, cnt)
    FROM docs
    CLUSTER BY word
) a
REDUCE word, cnt USING 'python wc_reduce.py';
```

Fig. 3.2 An example HiveQl query

(e.g., tables, columns, partitions) and a subset of SQL to the unstructured world of Hadoop while still maintaining the extensibility and flexibility that Hadoop provides. Thus, it supports all the major primitive types (e.g., integers, floats, strings) as well as complex types (e.g., maps, lists, structs). Hive supports queries expressed in an SQL-like declarative language, *HiveQL*,[2] and, therefore, can be easily understood by anyone who is familiar with SQL. These queries are compiled into MapReduce jobs that are executed using Hadoop. In addition, HiveQL enables users to plug in custom MapReduce scripts into queries [64]. For example, the canonical MapReduce word count example on a table of documents (Fig. 2.1) can be expressed in HiveQL as depicted in Fig. 3.2 where the *MAP* clause indicates how the input columns (*doctext*) can be transformed using a user program ("python wc_mapper.py") into output columns (*word* and *cnt*). The *REDUCE* clause specifies the user program to invoke ("python wc_reduce.py") on the output columns of the subquery.

HiveQL supports Data Definition Language (DDL) statements which can be used to create, drop, and alter tables in a database. It allows users to load data from external sources and insert query results into Hive tables via the load and insert Data Manipulation Language (DML) statements, respectively. However, HiveQL currently does not support the update and deletion of rows in existing tables (in particular, INSERT INTO, UPDATE, and DELETE statements) which allows the use of very simple mechanisms to deal with concurrent read and write operations without implementing complex locking protocols. The metastore component is the Hive's system catalog which stores metadata about the underlying table. This metadata is specified during table creation and reused every time the table is referenced in HiveQL.

Hive is able to run on top of Hadoop, Spark, or Tez [65] as its execution engine. These execution engines run on top of the YARN framework [66] for task management, resource allocation, and memory management. HiveQL queries are implicitly converted into tasks which are then executed by the underlying engine under YARN management. The results are then collected by YARN and sent to the

---

[2]http://wiki.apache.org/hadoop/Hive/LanguageManual.

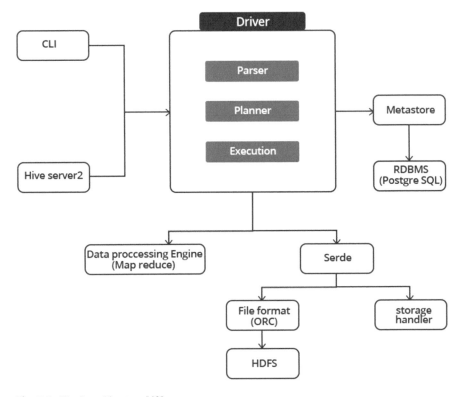

Fig. 3.3 Hive's architecture [42]

Hive client. Figure 3.3 illustrates the main components of Hive Architecture which are described as follows:

- *Metastore*: This component is responsible for storing the schema information and the tables that are maintained in Hive. It also stores the locations of table files.
- *Driver*: This node of Hive cluster controls the execution of the HiveQL statement. It starts the execution of the statement, collects and stores intermediate and final results, and ends the execution of the statement.
- *Compiler*: This component compiles the HiveQL query and generates the query execution plan. The execution plan contains the stages and jobs to be executed by the underlying execution engine to get the desired output of the query.
- *Optimizer*: This component performs the required transformations on the execution plan in order to find the most efficient physical query execution plan.
- *Executor*: Executes the task and interacts with the Hadoop Job Tracker to control task execution.
- *User Interface*: Hive CLI provides an interface for connecting to Hive, submitting queries, and monitoring processes. Hive also provides a Thrift server which supports Thrift protocol for external clients to communicate with Hive.

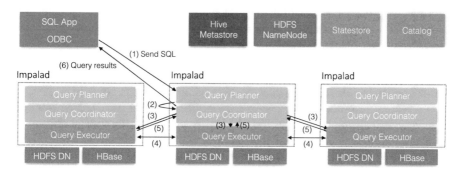

Fig. 3.4 Impala's architecture [67]

3.3 Impala

Apache Impala[3] is another open-source project, built by Cloudera, to provide a massively parallel processing SQL query engine that runs natively in Apache Hadoop [67]. It utilizes the standard components of Hadoop's infrastructure (e.g., HDFS, HBase, YARN) and is able to read the majority of the widely used file formats (e.g., Parquet,[4] Avro,[5] RCFile[6]). Therefore, by using Impala, the user can query data which is stored in Hadoop Distributed File System (HDFS). It also uses the same metadata, SQL syntax (HiveQL), that Apache Hive uses. However, Impala does not use the Hadoop execution engine to run the queries. Instead, it relies on its own set of daemons, which are installed alongside the datanodes and are tuned to optimize local processing and to avoid bottlenecks.

Figure 3.4 illustrates the architecture of the Impala system that consists of three main components: The *Impala daemon (impalad)* that accepts queries from client processes and orchestrates their execution across the cluster. The Impala daemon that operates in the first role by managing query execution is considered the coordinator for that query. However, all Impala daemons are symmetric. That means they may all operate in all roles which help with fault tolerance and with load balancing. The *Statestore daemon (statestored)* is a metadata publish-subscribe component which disseminates cluster-wide metadata to all Impala processes. *The Catalog daemon (catalogd)* serves as a catalog and metadata access repository and is responsible for broadcasting any changes to the system catalog as well. In general, the gathering of metadata and table statistics in a Hadoop environment is known to be complicated because new data can show up simply by moving

---

[3]https://www.cloudera.com/products/open-source/apache-hadoop/impala.html.

[4]https://parquet.apache.org/.

[5]https://avro.apache.org/.

[6]https://hive.apache.org/javadocs/r0.12.0/api/org/apache/hadoop/hive/ql/io/RCFile.html.

data files into a table's root directory. Impala tackles this problem by detecting new data files automatically via running a background process which also updates the metadata and schedules queries which compute the incremental table statistics. Impala is currently restricted to relational schemas which is often adequate for pre-existing data warehouse workloads; however, it does not yet support file formats that allow nested relational schemas or other complex column types (e.g., structs, arrays, maps). In order to improve execution times, Impala uses LLVM runtime code generation using techniques [68]. LLVM is a compiler library and collection of related tools which is designed to be modular and reusable. It allows applications like Impala to perform compilation within a running process, with the full benefits of a modern optimizer and the ability to generate machine code for a number of architectures by exposing separate APIs for all steps of the compilation process.

3.4   IBM Big SQL

Big SQL[7] is the SQL interface for the IBM big data processing platform, InfoSphere BigInsights,[8] which is built on the Apache Hadoop framework. In particular, it provides SQL access to data that is stored in InfoSphere BigInsights and uses the Hadoop framework for complex datasets and direct access for smaller queries. In the initial implementation of Big SQL, the engine was designed to decompose the SQL query into a series of Hadoop jobs. For interactive queries, Big SQL relied on a built-in query optimizer that re-writes the input query as a local job to help minimize latencies by using Hadoop dynamic scheduling mechanisms. The query optimizer also takes care of traditional query optimization such as optimal order, in which tables are accessed in the order where the most efficient join strategy is implemented for the query. The design of the recent version of the Big SQL engine has been implemented by adopting a shared-nothing parallel database architecture, in which it replaces the underlying Hadoop framework with a massively parallel processing SQL engine that is deployed directly on the physical Hadoop Distributed File System (HDFS). Therefore, the data can be accessed by all other tools of the Hadoop ecosystem, such as Pig and Hive. The system infrastructure provides a logical view of the data through the storage and management of metadata information. In particular, a table is simply a view that is defined over the stored data in the underlying HDFS. In addition, the Big SQL engine uses the Apache Hive database catalog facility for storing the information about table definitions, location, and storage format. Figure 3.5 illustrates the architecture of IBM Big SQL engine.

---

[7]http://www-01.ibm.com/software/data/infosphere/hadoop/big-sql.html.
[8]https://www.ibmbigdatahub.com/tag/332.

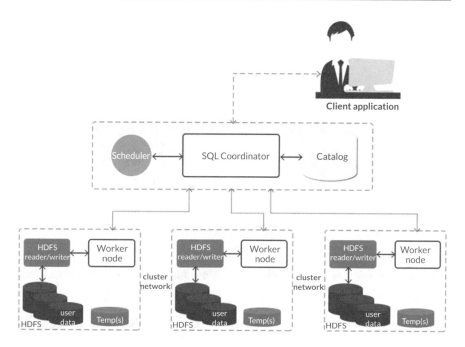

Fig. 3.5 The architecture of IBM Big SQL engine

3.5 SPARK SQL

SparkSQL [49] has been introduced as an alternative interface for Spark that integrates relational processing with Spark's functional programming API [44]. In particular, SparkSQL bridges the gap between the two models by providing a *DataFrame* API that can execute relational operations on both external data sources and Spark's built-in distributed collections [49]. It provides three main capabilities for interfacing with structured and semi-structured data:

- It provides the *Data Frame* abstraction which is similar to the table abstraction in relational databases and simplifies working with structured data.
- It is able to read and write datasets in different structured formats including JSON, Hive Tables, and Parquet file format.[9]
- It provides an SQL interface for interacting with the datasets stored in the Spark engine. The SQL interface is available via the command-line or over JDBC-ODBC connectors.

---

[9]https://parquet.apache.org/.

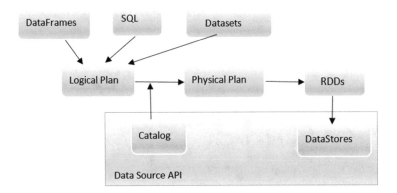

Fig. 3.6 Architecture of Spark SQL

Figure 3.6 illustrates the architecture of Spark SQL. It uses a component called the Catalyst Optimizer for the generation and optimization of logical query plan. Spark SQL consists of the following main layers:

- A Language API which provides support for standard SQL and HiveQL queries.
- Schema RDD which enables Spark to work with schemas tables and records.
- Data Sources which allow Spark to work with other data stores apart from text files and Avro files.[10]

SparkSQL relies on an extensible optimizer, called Catalyst, which supports adding data sources, optimization rules, and data types for domains such as machine learning. In SparkSQL, DataFrames are collections of structured records that can be manipulated using Spark's procedural API, or using new relational APIs that allow richer optimizations. DataFrames can be instantiated directly using Spark's built-in distributed collections of Java/Python objects, enabling relational processing in existing Spark programs. Catalyst uses features of the Scala programming language including pattern-matching for expressing composable rules in a Turing-complete language. Catalyst offers a general framework for transforming trees which is used to perform analysis, planning, and runtime code generation. In addition, Catalyst can be extended with new data sources (e.g., JSON) to which filters can be pushed; with user-defined functions; and with user-defined types for domains such as machine learning [49, 50].

---

[10]https://avro.apache.org/.

3.6   HadoopDB

There has been a long debate on the comparison between MapReduce framework and parallel database systems [70]. Pavlo et al. [19] have conducted a large-scale comparison between the Hadoop implementation of MapReduce framework and parallel SQL database management systems in terms of performance and development complexity. The results of this comparison have shown that parallel database systems displayed a significant performance advantage over MapReduce in executing a variety of data-intensive analysis tasks. On the other hand, the Hadoop implementation was very much easier and more straightforward to set up and use in comparison to that of the parallel database systems. MapReduce have also shown to have superior performance in minimizing the amount of work that is lost when a hardware failure occurs. In addition, MapReduce (with its open-source implementations) represents a very cheap solution in comparison to the very financially expensive parallel DBMS solutions (the price of an installation of a parallel DBMS cluster usually consists of seven figures of US Dollars) [70].

The *HadoopDB* project[11] is a hybrid system which is designed to attempt combining the scalability advantages of Hadoop framework with the performance and efficiency merits of parallel databases [69]. The basic idea behind HadoopDB is to cluster multiple single-node database systems (PostgreSQL) using Hadoop as the task coordinator and network communication layer. Queries are expressed in SQL but their execution is parallelized across nodes using the MapReduce framework and as much as possible is pushed inside of the corresponding node databases. Thus, HadoopDB tries to achieve fault tolerance and the ability to operate in heterogeneous environments by inheriting the scheduling and job tracking implementation from Hadoop. Parallelly, it tries to achieve the performance of parallel databases by doing most of the query processing inside the database engine. Figure 3.7 illustrates the architecture of HadoopDB which consists of two layers:

- A data storage layer or the Hadoop Distributed File System (HDFS).
- A data processing layer or the MapReduce framework.

In this architecture, HDFS is a block-structured file system managed by a central *NameNode*. Individual files are broken into blocks of a fixed size and distributed across multiple *DataNodes* in the cluster. The NameNode maintains metadata about the size and location of blocks and their replicas. The MapReduce framework follows a simple master–slave architecture. The master is a single *JobTracker* and the slaves or worker nodes are *TaskTrackers*. The JobTracker handles the runtime scheduling of MapReduce jobs and maintains information on each TaskTracker's load and available resources. The *Database Connector* is the interface between independent database systems residing on nodes in the cluster and TaskTrackers. The Connector connects to the database, executes the SQL query, and returns results

---

[11]http://db.cs.yale.edu/hadoopdb/hadoopdb.html.

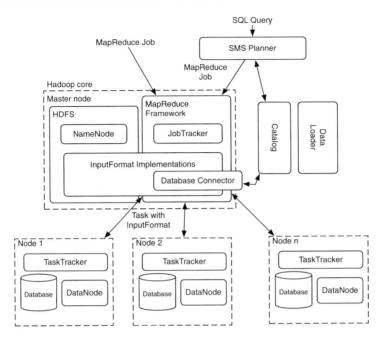

Fig. 3.7 HadoopDB's architecture [69]

as key-value pairs. The *Catalog* component maintains metadata about the databases, their location, replica locations, and data partitioning properties. The *Data Loader* component is responsible for globally repartitioning data on a given partition key upon loading and breaking apart single-node data into multiple smaller partitions or chunks. The *SMS planner* extends the HiveQL translator [63] and transforms SQL into MapReduce jobs that connect to tables stored as files in HDFS.

3.7   Presto

Facebook has released *Presto*[12] as an open-source distributed SQL query engine for running interactive analytic queries against large-scale structured data sources of sizes of gigabytes up to petabytes. In particular, it targets analytic operations where expected response times range from sub-second to minutes. Presto allows querying data where it lives, including Hive, NoSQL databases (e.g., Cassandra[13]), relational databases, or even proprietary data stores. Therefore, a single Presto query can combine data from multiple sources. Presto is offered as a cloud service by

---

[12]http://prestodb.io/.

[13]http://cassandra.apache.org/.

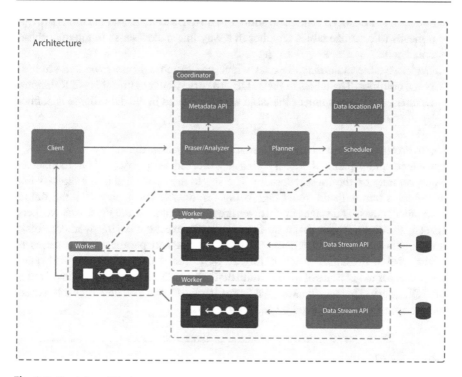

Fig. 3.8 Presto's architecture

Qubole.[14] Presto has been recently adopted by big companies and applications such as Netflix[15] and Airbnb.[16]

Figure 3.8 illustrates the architecture of PrestoDB with the following key components:

- *Connector*: A connector enables PrestoDB to communicate with a data source. Connectors can be considered as the data source drivers. PrestoDB comes with predefined connectors such as Hive and MySQL. There are also several connectors, built by third-party developers, which enable it to connect to other data sources. Every catalog in PrestoDB must be associated with a connector.
- *Catalog*: This component contains the data schemas and a reference to a connector. Catalogs are defined in properties files stored in PrestoDB configuration directory. Catalog name is the first part of the fully qualified name of a PrestoDB table.

---

[14]https://www.qubole.com/presto-as-a-service/.

[15]http://techblog.netflix.com/2014/10/using-presto-in-our-big-data-platform.html.

[16]https://gigaom.com/2015/03/05/airbnb-open-sources-sql-tool-built-on-facebooks-presto-database/.

- *Schema*: A PrestoDB schema is similar to a RDBMS database. Schemas provide a means to organize tables together in a way that makes sense to the underlying data source.
- *Table*: A table maintains the data which are structured into rows and strongly typed columns. The tables in PrestoDB are very similar to the tables of Relational Databases. The mapping of the table which is stored in the data source is defined by the connector.

When PrestoDB receives an SQL statement, it parses the query and creates a distributed query plan. The query plan is executed as a series of interconnected stages running on the Presto workers. The stages are connected in a hierarchy that resembles a tree with the root stage being responsible for aggregating the output from other stages. The stages themselves are not executed on the PrestoDB workers. Instead, the tasks which make up the stages are executed on the workers. Tasks have inputs and outputs, and they can be executed in parallel with a series of drivers. When PrestoDB is scheduling an SQL query, the coordinator will query the connector to get a list of all the splits that are available for the tables involved in the SQL query. The coordinator then keeps track of tasks executed by each worker and the splits being processed by each of the tasks.

3.8　Tajo

Apache Tajo[17] is another distributed data warehouse system for Apache Hadoop that is designed for low-latency and scalable ad-hoc queries ETL processes. Tajo can analyze data which is stored on HDFS, Amazon S3, OpenStack Swift[18] and local file systems. It provides an extensible query re-write system that lets users and external programs query data through SQL. In Tajo, a distributed execution plan is represented as a direct acyclic graph (DAG) and it uses cost-based join order and a rule-based optimizer [71]. Tajo does not use MapReduce and has its own distributed execution framework.

　Figure 3.9 illustrates the architecture of the Tajo system where a Tajo cluster instance consists of one *TajoMaster* and a number of *TajoWorkers*. The TajoMaster coordinates the cluster membership and their resources. In addition, it processes datasets stored in the underlying storage instances. When a user submits an SQL query, TajoMaster decides whether the query is immediately executed only in TajoMaster or the query is executed across a number of TajoWorkers. Depending on the decision, the TajoMaster either forwards the query to workers or it does not.

---

[17]http://tajo.apache.org/.

[18]http://docs.openstack.org/developer/swift/.

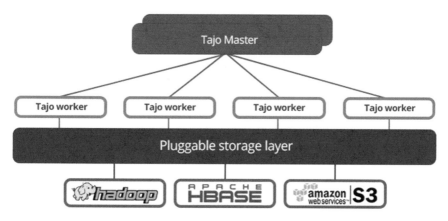

Fig. 3.9 Tajo's architecture

3.9 Google Big Query

Google has introduced the *Dremel* project [72], open-sourced as *Apache Drill*[19] and commercialized as *BigQuery*,[20] as a scalable, interactive ad-hoc query system for the analysis of read-only nested data. It represents a Schema-free SQL engine that combines multi-level execution trees and columnar data layout over thousands of CPUs and petabytes of data. Dremel provides a high-level, SQL-like language to express ad-hoc queries which are executed natively without translating them into Hadoop jobs. Dremel uses a column-striped storage representation, which enables it to read less data from secondary storage and reduce CPU cost due to cheaper compression. Therefore, it is able to support schema-based formats such as Protocol (e.g., Trevni,[21] CSV) and schemaless formats (e.g., JSON, BSON). Apache Drill 1.0 is now shipping with the MapR Apache Hadoop distribution.

3.10 Phoenix

Apache Phoenix[22] is a project which started as an internal project by Salesforce[23] before it was open-sourced. It provides a Java layer that enables the developers to execute SQL queries on top of Apache HBase. Phoenix supports SQL syntax which allows the inputs and outputs to be represented using standard JDBC APIs instead

---

[19]http://drill.apache.org/.

[20]https://cloud.google.com/bigquery/.

[21]https://avro.apache.org/docs/1.7.7/trevni/spec.html.

[22]https://phoenix.apache.org/.

[23]http://www.salesforce.com/.

of HBase's Java client APIs. In particular, the input SQL query is compiled it into a series of native HBase API calls which pushes as much work as possible into the cluster for parallel execution. Unlike most of the other systems, Phoenix is intended to operate *exclusively* on the HBase stored data so that its design and implementation are heavily customized to leverage HBase features including coprocessors and skip scans.

3.11   Polybase

Microsoft has developed the *Polybase*[24] project which allows SQL Server Parallel Data Warehouse (PDW) users to execute queries against data stored in Hadoop, specifically the HDFS. Polybase provides users with the ability to move data in parallel between nodes of the Hadoop and PDW clusters. In addition, users can create external tables over HDFS-resident data. This allows queries to reference data stored in HDFS as if it were loaded into a relational table. In addition, users can seamlessly perform joins between tables in PDW and data in HDFS. When optimizing an SQL query that references data stored in HDFS, the Polybase query optimizer makes a cost-based decision (using statistics on the HDFS file stored in the PDW catalog) on whether or not it should transform relational operators over HDFS-resident data into Hadoop jobs which are then sent for execution using the underlying Hadoop cluster [73]. In addition, Polybase is capable of fully leveraging the larger compute and I/O power of the Hadoop cluster by moving work to Hadoop for processing even for queries that only reference PDW-resident data [74].

---

[24]http://gsl.azurewebsites.net/Projects/Polybase.aspx.

Large-Scale Graph Processing Systems

4

4.1 The Challenges of Big Graphs

Recently, people, devices, processes, and other entities have been more connected than at any other point in history. In general, graph is a natural, neat, and flexible structure to model the complex relationships, interactions, and interdependencies between objects (Fig. 4.1). In particular, each graph consists of nodes (or vertices) that represent objects and edges (or links) that represent the relationships among the graph nodes. Graphs have been widely used to represent datasets in a wide range of application domains (Fig. 4.2) such as social science, astronomy, computational biology, telecommunications, computer networks, semantic web, protein networks, and many more [75].

The web graph is a dramatic example of a large-scale graph. Google estimates that the total number of web pages has exceeded 1 trillion; experimental graphs of the World Wide Web contain more than 20 billion nodes (pages) and 160 billion edges (hyperlinks). Graphs of social networks are another example. For instance, Facebook went from roughly 1 million users in 2004 to 2.5 billions in 2019.[1] In the Semantic Web context, the ontology of DBpedia (derived from Wikipedia) contains 3.7 million objects (nodes) and 400 million facts (edges).

The ever-increasing size of graph-structured data for these applications creates a critical need for scalable systems that can process large amounts of it efficiently. In practice, graph analytics is an important and effective big data discovery tool. For example, it enables identifying influential persons in a social network, inspecting fraud operations in a complex interaction network, and recognizing product affinities by analyzing community buying patterns. However, with the enormous growth in graph sizes, huge amounts of computational power would be

---

[1]https://www.statista.com/statistics/264810/number-of-monthly-active-facebook-users-worldwide/.

S. Sakr, *Big Data 2.0 Processing Systems*,
https://doi.org/10.1007/978-3-030-44187-6_4

Fig. 4.1 Graph-based modeling

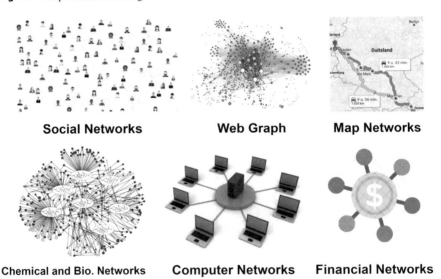

Fig. 4.2 Graph-based applications

required to analyze such massive graph datasets. In practice, distributed and scalable processing of massive graph datasets is a challenging task and has its own specific challenges on top of the general challenges of the big data processing problems [76].

For instance, the iterative nature of graph processing and computation algorithms typically involves extensive communication and message passing between the graph nodes in each processing step. In addition, graph algorithms tend to be explorative with random access patterns which are commonly challenging to predict. Furthermore, due to their inherent irregular structure, dealing with the graph partitioning represents a fundamental challenge due to its direct impact on load balancing among the processing nodes, communication cost, and consequently the whole system performance [77].

4.2   Does Hadoop Work Well for Big Graphs?

In principle, general-purpose distributed data processing frameworks such as Hadoop are well suited for analyzing unstructured and tabular data. However, such frameworks are not efficient for directly implementing iterative graph algorithms which often require multiple stages of complex joins [22]. In addition, the general-purpose join and aggregation mechanisms defined in such distributed frameworks are not designed to leverage the common patterns and structure in iterative graph algorithms. Therefore, such obliviousness of the graph structure leads to huge network traffic and missed opportunities to fully leverage important graph-aware optimizations.

In practice, a fundamental aspect in the design of the Hadoop framework is that it requires the results of each single map or reduce task to be *materialized* into a local file on the Hadoop Distributed File System (HDFS), a distributed file system to store data files across multiple machines, before it can be processed by the following tasks (Fig. 4.3). This materialization step supports the implementation of

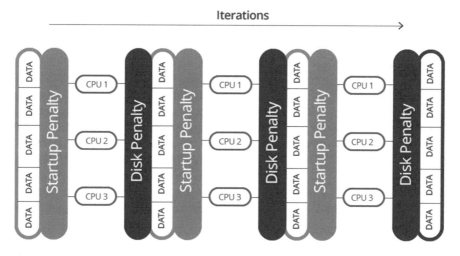

Fig. 4.3 MapReduce iteration

a simple and elegant checkpointing/restarting fault tolerance mechanism. However, from the graph processing point of view, this materialization step dramatically hurts the performance of graph processing algorithms (e.g., PageRank, Connected Component, Triangle Count) which are iterative in nature and typically aim to traverse the graph in specific way. Hence, in practice, the efficiency of graph computations depends heavily on inter-processor bandwidth as graph structures are sent over the network iteration after iteration. In addition, the MapReduce framework does not directly support iterative data analysis applications. To implement iterative programs, programmers might manually issue multiple MapReduce jobs and orchestrate their execution with a driver program. In practice, the manual orchestration of an iterative program in MapReduce has two key problems:

- While much data might be unchanged from iteration to iteration, the data must be reloaded and reprocessed at each iteration, wasting I/O, network bandwidth, and processor resources.
- The termination condition might involve the detection of when a fix point is reached. The condition itself might require an extra MapReduce job on each iteration, again increasing resource use in terms of scheduling extra tasks, reading extra data from disk, and moving data across the network.

Several approaches have proposed Hadoop extensions (e.g., *HaLoop* [27], *Twister* [29], *iMapReduce* [28]) to optimize the iterative support of the MapReduce framework. For instance, the *HaLoop* system [27] has been designed to support iterative processing on the MapReduce framework by extending the basic MapReduce framework with two main functionalities: Caching the invariant data in the first iteration and then reusing them in later iterations in addition to caching the reducer outputs, which makes checking for a fix point more efficient, without an extra MapReduce job. The *iMapReduce* framework [28] supports the feature of iterative processing by keeping alive each map and reduce task during the whole iterative process. In particular, when all of the input data of a persistent task are parsed and processed, the task becomes dormant, waiting for the new updated input data. For a map task, it waits for the results from the reduce tasks and is activated to work on the new input records when the required data from the reduce tasks arrive. For the reduce tasks, they wait for the map tasks' output and are activated synchronously as in MapReduce. *Twister*[2] is a MapReduce runtime with an extended programming model that supports iterative MapReduce computations efficiently [29]. It uses a publish/subscribe messaging infrastructure for communication and data transfers, and supports long running map/reduce tasks. In particular, it provides programming extensions to MapReduce with broadcast and scatter type data transfers. Microsoft has also developed a project that provides an iterative MapReduce runtime for Windows Azure called *Daytona*.[3] In general, these approaches remain inefficient

---

[2]http://www.iterativemapreduce.org/.
[3]http://research.microsoft.com/en-us/projects/daytona/.

for the graph processing case because the efficiency of graph computations depends heavily on inter-processor bandwidth as graph structures are sent over the network after each iteration. While most of the data might be unchanged from iteration to iteration, the data must be reloaded and reprocessed at each iteration, resulting in the unnecessary wastage of I/O, network bandwidth, and processor resources. In addition, the termination condition might involve the detection of when a fix point is reached. The condition itself might require an extra MapReduce task on each iteration, again increasing the resource usage in terms of scheduling extra tasks, reading extra data from disk, and moving data across the network.

Other approaches have attempted to implement graph processing operations on top of the MapReduce framework (e.g., *Surfer* [78], *PEGASUS* [79]). For example, the *Surfer* system [78] has been presented as a large-scale graph processing engine which is designed to provide two basic primitives for programmers: MapReduce and propagation. In this engine, MapReduce processes different key-value pairs in parallel, and propagation is an iterative computational pattern that transfers information along the edges from a vertex to its neighbors in the graph. In particular, to use the graph propagation feature in the Surfer system, the user needs to define two functions: transfer and combine. The transfer function is responsible for exporting the information from a vertex to its neighbors, while the combine function is responsible for aggregating the received information at each vertex. In addition, the Surfer system adopts a graph partitioning strategy that attempts to divide the large graph into many partitions of similar sizes so that each machine can hold a number of graph partitions and manage the propagation process locally before exchanging messages and communicating with other machines. As a result, the propagation process can exploit the locality of graph partitions for minimizing the network traffic. *GBASE*[4] is another MapReduce-based system that uses a graph storage method, called *block compression*, which first partitions the input graph into a number of blocks [80]. According to the partition results, GBASE reshuffles the nodes so that the nodes belonging to the same partition are placed near to each other after which it compresses all non-empty blocks through a standard compression mechanism such as *GZip*.[5] Finally, it stores the compressed blocks together with some meta information into the graph storage. GBASE supports different types of graph queries including *neighborhood*, *induced subgraph*, *EgoNet*, *K-core*, and *cross-edges*. To achieve this goal, GBASE applies a grid selection strategy to minimize disk accesses and answer queries by applying a MapReduce-based algorithm that supports incidence matrix-based queries. Finally, *PEGASUS*[6] is a large-scale graph mining library that has been implemented on top of the Hadoop framework and supports performing typical graph mining tasks such as *computing the diameter of the graph, computing the radius of each node* and *finding the connected components* via using Generalized Iterative Matrix-Vector multiplication (GIM-V)

---

[4]http://systemg.research.ibm.com/analytics-search-gbase.html.

[5]http://www.gzip.org/.

[6]http://www.cs.cmu.edu/~pegasus/.

which represents a generalization of normal matrix–vector multiplication [79, 81]. The library has been utilized for implementing a MapReduce-based algorithm for discovering patterns on near-cliques and triangles on large-scale graphs [82]. In practice, GBASE and PEGASUS are unlikely to be intuitive for most developers, who might find it challenging to think of graph processing in terms of matrices. Also, each iteration is scheduled as a separate Hadoop job with increased workload: When the graph structure is read from disk, the map output is spilled to disk and the intermediary result is written to the HDFS.

To solve this inherent performance problem of the MapReduce framework, several specialized platforms which are designed to serve the unique processing requirements of large-scale graph processing have recently emerged. These systems provide programmatic abstractions for performing iterative parallel analysis of large graphs on clustered systems. In particular, in 2010, Google has pioneered this area by introducing the *Pregel* [43] system as a scalable platform for implementing graph algorithms. Since then, we have been witnessing the development of a large number of scalable graph processing platforms. For example, the Pregel system has been cloned by various open-source projects such as *Apache Giraph*[7] and *Apache Hama*.[8] It has also been further optimized by other systems such as *Pregelix* [61], *Mizan* [83], and *GPS* [84]. In addition, a family of related systems [85–87] has been initiated by the *GraphLab* system [87] as an open-source project at Carnegie Mellon University that are now supported by Dato Inc.[9] Furthermore, some other systems have also been introduced such as *GraphX* [51], *Trinity* [88], *GRACE* [89], and *Signal/Collect* [90] (Fig. 4.4).

4.3   Pregel Family of Systems

4.3.1   The Original Architecture

Bulk synchronous parallel (BSP) is a parallel programming model that uses a message passing interface (MPI) to address the scalability challenge of parallelizing jobs across multiple nodes [91]. In principle, BSP is a vertex-centric programming model where the computation on vertices are represented as a sequence of *super-steps* (iterations) with synchronization between the nodes participating at superstep barriers and each vertex can be active or inactive at each iteration (superstep) (Fig. 4.5). Such a programming model can be seen as a graph extension of the actors programming model [92] where each vertex represents an actor and edges represent the communication channel between actors. In such model, users can focus on specifying the computation on the graph nodes and the communication

---

[7]http://giraph.apache.org/.

[8]http://hama.apache.org/.

[9]https://dato.com/.

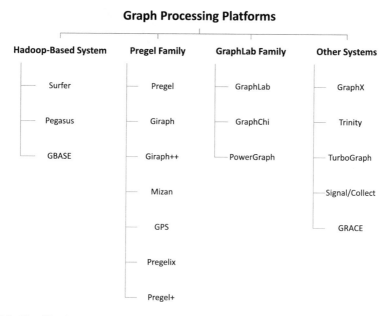

Fig. 4.4 Classification of graph processing platforms

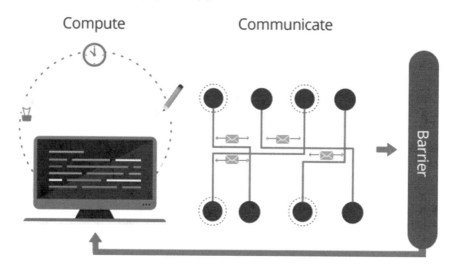

Fig. 4.5 BSP programming model

among them without worrying about the specifics of the underlying organization or resource allocation of the graph data.

In 2010, the *Pregel* system [43], presented by Google and implemented in C/C++, has been introduced as the first BSP implementations that provides a native API specifically for programming graph algorithms using a "*think like a*

Fig. 4.6   Vertex-based programming model

vertex" computing paradigm where each vertex has a value and can exchange message with other graph vertices in a number of iterations. During each iteration of the computation, each vertex can receive messages, updates its value, and send messages to its neighbor vertices (Fig. 4.6). In order to avoid communication overheads, Pregel preserves data locality by ensuring computation is performed on locally stored data. In particular, Pregel distributes the graph vertices to the different machines of the cluster where each vertex and its associated set of neighbors are assigned to the same node. Graph processing algorithms are then represented as supersteps where each step defines what each participating vertex has to compute and edges between vertices represent communication channels for transmitting computation results from one vertex to another. In particular, at each superstep, a vertex can execute a user-defined function, send or receive messages to its neighbors (or any other vertex with a known ID), and change its state from active to inactive. Pregel supersteps are synchronous, that means each superstep is concluded once all active vertices of this steps have completed their computations and all of the exchanged messages among the graph vertices have been delivered. Pregel can start the execution of a new superstep $(S + 1)$ only after the current superstep (S) completes its execution. Each superstep ends with a waiting phase, synchronization barrier (Fig. 4.7), that ensures that messages sent from one superstep are correctly delivered to the subsequent step. In each superstep, a vertex may vote to halt (inactive status) if it does not receive any message and it can also be reactivated once it receives a message at any subsequent superstep (Fig. 4.8). Thus, in each superstep, only the active vertices are involved in the computation process which results in significant reduction in the communication overhead (Fig. 4.9), a main advantage against the Hadoop-based processing for graphs. The whole graph processing operation terminates when all vertices are inactive and no more messages are in transit between the vertices of the graph. In Pregel, the input graph is loaded once at the beginning of the program and all computations are executed in-memory. Pregel uses a master/workers model where the master node is responsible for coordinating synchronization at the superstep barriers while each worker independently invokes and executes the `compute()` function on the vertices of its assigned portion of the graph and maintains the message queue to receive messages from the vertices of other workers.

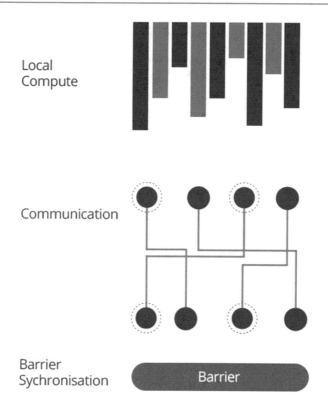

Local
Compute

Communication

Barrier
Sychronisation

Fig. 4.7 Supersteps execution in BSP

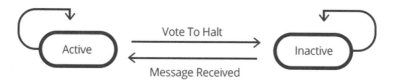

Fig. 4.8 Vertex voting in BSP

4.3.2   Giraph: BSP + Hadoop for Graph Processing

In 2012, *Apache Giraph* has been introduced as an open-source project that clones the ideas and implementation of the Pregel specification in Java on top of the infrastructure of the Hadoop framework. In principle, the relationship between the Pregel system and Giraph project is similar to the relationship between the MapReduce framework and the Hadoop project (Fig. 4.10). Giraph has been initially implemented by Yahoo!. Later, Facebook built its Graph Search services using Giraph. Currently, Giraph enlists contributors from Yahoo!, Facebook, Twitter, and LinkedIn. Giraph runs graph processing jobs as map-only jobs on Hadoop and

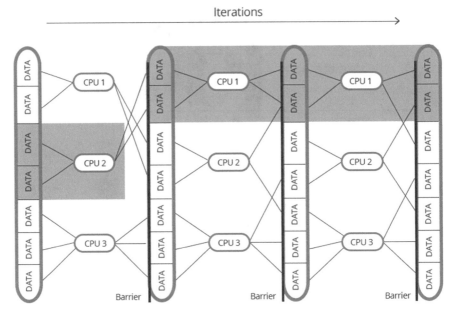

Fig. 4.9 BSP execution for iterative processing

Google MapReduce   Google Pregel

Fig. 4.10 Pregel and Giraph

uses HDFS for data input and output. Giraph also uses *Apache ZooKeeper*[10] for coordination, checkpointing, and failure recovery schemes.

---

[10]http://zookeeper.apache.org/.

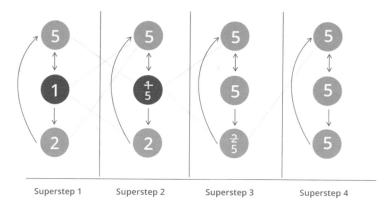

Superstep 1	Superstep 2	Superstep 3	Superstep 4

Fig. 4.11 BSP example

Primarily, Giraph provides a distributed execution engine for graph computations. In addition, it supports an API-based programming model that allows its
user to focus on designing the logic of their graph computation without the need to
worry about how the underlying graph is stored in disk or loaded in memory or how
the execution of the computation is distributed among the nodes of the computing
cluster with fault tolerance considerations. In Giraph, graph processing programs
are expressed as a sequence of iterations (supersteps). During a superstep, the
framework starts a user-defined function for each vertex, conceptually in parallel.
The user-defined function specifies the behavior at a single vertex V and a single
superstep S. The function can read messages that are sent to V in superstep $S - 1$,
send messages to other vertices that are received at superstep $S + 1$, and modify the
state of V and its outgoing edges. Messages are typically sent along outgoing edges,
but you can send a message to any vertex with a known identifier. Each superstep
represents atomic units of parallel computation.

Figure 4.11 illustrates an example of the communicated messages between a
set of graph vertices for computing the maximum vertex value. In this example,
in Superstep 1, each vertex sends its value to its neighbor vertex. In Superstep 2,
each vertex compares its value with the received value from its neighbor vertex. If
the received value is higher than the vertex value, then it updates its value with the
higher value and sends the new value to its neighbor vertex. If the received value is
lower than the vertex value, then the vertex keeps its current value and votes to halt.
Hence, in Superstep 2, only the vertex with value 1 updates its value to the higher
received value (5) and sends its new value. This process happens again in Superstep
3 for the vertex with the value 2, while in Superstep 4 all vertices vote to halt and
the program ends.

Similar to the Hadoop framework, Giraph is an efficient, scalable, and fault-
tolerant implementation on clusters of thousands of commodity computers, with
the distribution-related details hidden behind an abstraction. On a machine that
performs computation, it keeps vertices and edges in memory and uses network

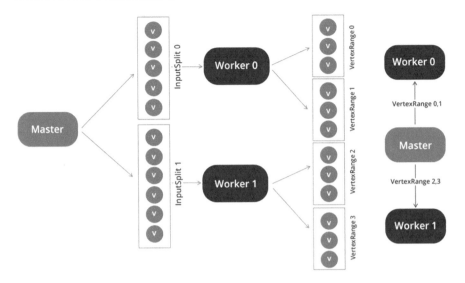

Fig. 4.12 Execution architecture of Giraph

transfers only for messages. The model is well suited for distributed implementations because it does not show any mechanism for detecting the order of execution within a superstep, and all communication is from superstep S to superstep $S + 1$. During program execution, graph vertices are partitioned and assigned to workers. The default partition mechanism is hash partitioning, but other custom partitioning schemes are also supported. Giraph applies a master/worker architecture, illustrated in Figure (Fig. 4.12). The master node assigns partitions to workers, coordinates synchronization, requests checkpoints, and collects health statuses. Similar to Hadoop, Giraph uses Apache ZooKeeper for synchronization. Workers are responsible for vertices. A worker starts the `compute()` function for the active vertices. It also sends, receives, and assigns messages with other vertices. During execution, if a worker receives input that is not for its vertices, it passes it along.

4.3.3   Pregel Extensions

The introduction of Google's Pregel has triggered much interest in the field of large-scale graph data processing and inspired the development of several Pregel-based systems which have been attempting to exploit different optimization opportunities. For instance, GPS^{11} is another Java implementation of Google's Pregel which comes from Stanford InfoLab [84]. GPS extends the Pregel API to allow certain global computation tasks to be specified and run by a master worker. In particular,

---

[11]http://infolab.stanford.edu/gps/.

it provides an additional function, `master.compute()`, that provides access to all of the global aggregated values, the global values which are transparent to the vertices. The global aggregated values can be updated before they are broadcast to the workers. GPS also offers the Large Adjacency List Partitioning (LALP) mechanism as an optional performance optimization for algorithms that send the same message to all of its neighbors. In particular, LALP works by partitioning the adjacency lists of high-degree vertices across different workers. For each partition of the adjacency list of a high-degree vertex, a mirror of the vertex is created in the worker that keeps the partition. When a high-degree vertex broadcasts a message to its neighbors, at most one message is sent to its mirror at each machine. Then, the message is forwarded to all its neighbors in the partition of the adjacency list of the high-degree vertex. This mechanism works well for algorithms like PageRank, weakly connected components (WCC), and single source shortest path (SSSP) with unit edge weights but does not work well for some other algorithms like distributed minimal spanning tree construction (DMST). Furthermore, GPS applies a dynamic repartitioning strategy based on the graph processing workload in order to balance the workload among all workers and reduce the number of exchanged messages over the network. In particular, GPS exchanges vertices between workers based on the amount of data sent by each vertex.

Similar to GPS, *Mizan* is an open-source project developed in C++ by KAUST, in collaboration with IBM Research [83]. Mizan's dynamic repartitioning strategy is based on monitoring the runtime characteristics of the graph vertices (e.g., their execution time, and incoming and outgoing messages) and uses this information, at the end of every superstep, to construct a migration plan with the aims of minimizing the variations across workers by identifying which vertices to migrate and to where to migrate them. The migration plan is executed between supersteps transparently to avoid interfering with the user's computations. Mizan uses a distributed hash table (DHT) to maintain the location of vertices. The DHT is updated whenever a vertex moves between different workers as part of the migration plan to ensure that the messages in the next superstep can reach the new location of the vertex.

Pregelix[12] is a large-scale graph processing platform that applies set-oriented, iterative dataflow approach to implement the BSP-based Pregel programming model [61]. In particular, Pregelix treats the messages and vertex states in the graph computation as relational tuples with a well-defined schema and uses relational database-style query evaluation techniques to execute the graph computation. For example, Pregelix treats a message exchange as a join operation that is followed by a group-by operation that embeds functions which capture the semantics of the graph computation program. Therefore, Pregelix generates a set of alternative physical evaluation strategies for each graph computation program and uses a cost model to select the target execution plan among them. The execution engine of Pregelix is *Hyracks* [57], a general-purpose shared-nothing dataflow engine. Given a graph processing job, Pregelix first loads the input graph dataset (the initial Vertex relation)

---

[12]http://pregelix.ics.uci.edu/.

from a distributed file system, i.e., HDFS, into a Hyracks cluster and partitions it using a user-defined partitioning function across the worker machines. Pregelix leverages B-tree index structures from the Hyracks storage library to store partitions of Vertex on worker machines. During the supersteps, at each worker node, one (or more) local indexes are used to store one (or more) partitions of the Vertex relation. After the eventual completion of the overall graph computation, the partitioned Vertex relation is scanned and dumped back to HDFS.

Pregel+[13] is another Pregel-based project implemented in C/C++ with the aim of reducing the number of exchanged messages between the worker nodes using a mirroring mechanism. In particular, Pregel+ selects the vertices for mirroring based on a cost model that analyzes the trade-off between mirroring and message combining. *Giraph++* [93] has proposed a *"think like a graph"* programming paradigm that opens the partition structure to the users so that it can be utilized within a partition in order to bypass the heavy message passing or scheduling facilities. In particular, the graph-centric model can make use of off-the-shelf sequential graph algorithms in distributed computation, allowing asynchronous computation to accelerate convergence rates, and naturally support existing partition-aware parallel/distributed algorithms.

4.4   GraphLab Family of Systems

4.4.1   GraphLab

GraphLab [87] is an open-source large-scale graph processing project, implemented in C++, which started at CMU. Unlike Pregel, GraphLab relies on the shared memory abstraction and the GAS (Gather, Apply, Scatter) processing model which is similar to but also fundamentally different from the BSP model that is employed by Pregel. The GraphLab abstraction consists of three main parts: the *data graph*, the *update function* and, the *sync operation*. The data graph represents a user-modifiable program state that both stores the mutable user-defined data and encodes the sparse computational dependencies. The update function represents the user computation and operates on the data graph by transforming data in small overlapping contexts called *scopes*. In the GAS model, a vertex collects information about its neighborhood in the *Gather* phase, performs the computations in the *Apply* phase, and updates its adjacent vertices and edges in the *Scatter* phase. As a result, in GraphLab, graph vertices can directly *pull* their neighbors' data (via Gather) without the need to explicitly receive messages from those neighbors. In contrast, in the BSP model of Pregel, a vertex can learn its neighbors' values only via the messages that its neighbors *push* to it. GraphLab offers two execution modes: *synchronous* and *asynchronous*. Like BSP, the synchronous mode uses the notion of communication barriers, while the asynchronous mode does not support the notion

---

[13]http://www.cse.cuhk.edu.hk/pregelplus/.

of communication barriers or supersteps. It uses distributed locking to avoid conflicts and to maintain serializability. In particular, GraphLab automatically enforces serializability by preventing adjacent vertex programs from running concurrently by using a fine-grained locking protocol that requires sequentially grabbing locks on all neighboring vertices. For instance, it uses an *Edge Consistency Model* that allows two vertices to be simultaneously updated if they do not share an edge. In addition, it applies a *Graph Coloring* mechanism where two vertices can be assigned the same color if they do not share an edge. It is the job of the system scheduler to determine the order that vertices can be updated.

4.4.2 PowerGraph

In practice, high-degree vertices in power-law graphs results in imbalanced workload during the execution of graph computations. Therefore, another member of the GraphLab family of systems, *PowerGraph* [85], has been introduced to tackle this challenge. In particular, PowerGraph relies on a vertex-cut partitioning scheme (Fig. 4.13) that cuts the vertex set in such a way that the edges of a high-degree vertex are handled by multiple workers. Therefore, as a trade-off, vertices are replicated across workers, and communication among workers is required to guarantee that the vertex value on each replica remains consistent. PowerGraph eliminates the degree dependence of the vertex program by directly exploiting the GAS decomposition to factor vertex programs over edges. Therefore, it is able to retain the "think-like-a-vertex" programming style while distributing the computation of a single vertex program over the entire cluster. In principle, PowerGraph attempts to merge the best features from both Pregel and GraphLab. From GraphLab, PowerGraph inherits the data-graph and shared-memory view of computation, thus eliminating the need for users to specify the communication of information. From Pregel, PowerGraph borrows the commutative, associative gather concept. PowerGraph supports both the highly parallel bulk synchronous Pregel model of computation and the computationally efficient asynchronous GraphLab model of computation.

4.4.3 GraphChi

Another member of the GraphLab family of systems is *GraphChi* [86]. Unlike the other *distributed* members of the family, GraphChi,[14] implemented in C++, is a *centralized* system which is designed to process massive graphs that are maintained on the secondary storage of a single machine. In particular, GraphChi relies on a *Parallel Sliding Windows* (PSWs) mechanism for processing very large graphs from disk. PSW is designed to require only a very small number of non-sequential accesses to the disk, and thus it can perform well on both SSDs and traditional

---

[14]http://graphlab.org/projects/graphchi.html.

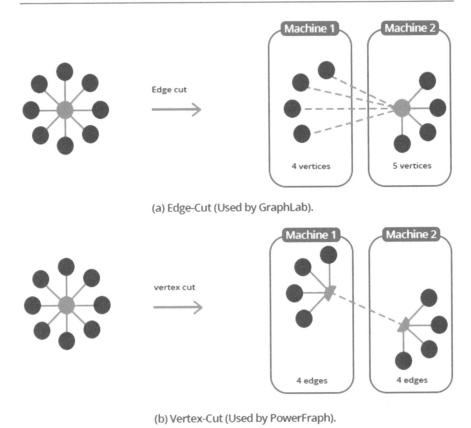

(a) Edge-Cut (Used by GraphLab).

(b) Vertex-Cut (Used by PowerFraph).

Fig. 4.13 Edge-cut vs vertex-cut partitioning schemes

hard drives. PSW partitions the input graph into subgraphs, called shards. In each shard, edges are sorted by the source IDs and loaded into memory sequentially. In addition, GraphChi supports a selective scheduling mechanism that attempts to converge faster on some parts of the graph especially on those where the change on values is significant. The main advantage of systems like GraphChi is that it avoids the challenge of finding efficient graph cuts that are balanced and can minimize the communication between the workers, which is a hard challenge. It also avoids other challenges of distributed systems such as cluster management and fault tolerance.

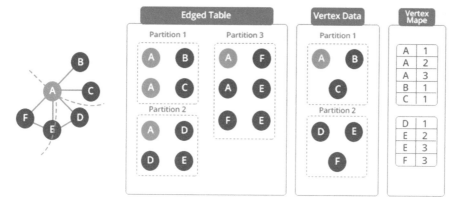

Fig. 4.14 Resilient distributed graph of GraphX

4.5   Spark-Based Large-Scale Graph Processing Systems

GraphX [51] is a distributed graph engine built on top of Spark. GraphX[15] extends Spark's Resilient Distributed Dataset (RDD) abstraction to introduce the resilient distributed graph (RDG), which associates records with vertices and edges in a graph and provides a collection of expressive computational primitives. The GraphX RDG leverages advances in distributed graph representation and exploits the graph structure to minimize communication and storage overhead. While the basic GraphX RDG interface naturally expresses graph transformations, filtering operations, and queries, it does not directly provide an API for recursive graph-parallel algorithms. Instead, the GraphX interface is designed to enable the construction of new graph-parallel APIs. In addition, unlike other graph processing systems, the GraphX API enables the composition of graphs with unstructured and tabular data and permits the same physical data to be viewed both as a graph and as collections without data movement or duplication. GraphX relies on a flexible vertex-cut partitioning to encode graphs as horizontally partitioned collections. By leveraging logical partitioning and lineage, GraphX achieves low-cost fault tolerance. In addition, by exploiting immutability, GraphX reuses indices across graph and collection views and over multiple iterations, reducing memory overhead and improving system performance.

A key difference of RDG from that of other graph systems (e.g., Pregel and GraphLab) is that RDG represents graph as records in tabular views of vertices and edges. In particular, as illustrated in Fig. 4.14, in RDG, a graph is represented as directed adjacency structure which is mapped into tabular records. Similar to RDDs, in RDGs, transformation from one graph to next creates another RDG with transformed vertices and edges via transformation operators. In addition,

---

[15]https://spark.apache.org/graphx/.

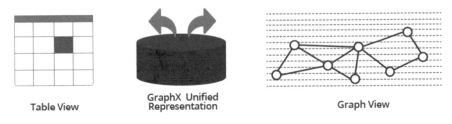

Fig. 4.15 Unified representation of graphs in GraphX

the GraphX API allows the same physical data to be viewed both as a graph and as collections without data movement or duplication (Fig. 4.15). GraphX also provides various operators for manipulating graphs (e.g., subgraph and mapVertices) and a library of common graph algorithms (e.g., PageRank and triangle counting). GraphX reuses indices across graph and collection views and over multiple iterations, reducing memory overhead and improving system performance. *GraphFrames* [94] is a graph package that has been designed on top of Spark's DataFrames. GraphFrames[16] benefits from the scalability and high performance of DataFrames. They provide a uniform API for graph processing available from Scala, Java, and Python. GraphFrames API implements DataFrame-based graph algorithms, and also incorporates simple graph pattern matching with fixed length patterns (called "motifs").

Recently, the *Morpheus* project[17] has been designed to enable the evaluation of Cypher queries over large property graphs using DataFrames on top of Apache Spark framework. In general, this framework enables combining the scalability of the Spark framework with the features and capabilities of Neo4j graph database by enabling the Cypher language to be integrated into the Spark analytics pipeline. Interestingly, graph processing and querying can be then easily interwoven with other Spark processing analytics libraries such as Spark GraphX, Spark ML, or Spark SQL. Moreover, this enables easy merging of graphs from Morpheus into Neo4j. Besides more advanced capabilities of Morpheus such as the ability to handle multiple graphs (i.e., graph composability) from different data sources even if they are not graph sources (i.e., relational data sources), it has the ability to create graph views on the data as well.

Figure 4.16 illustrates the architecture of Morpheus framework. In Morpheus, Cypher queries are translated into abstract syntax tree (AST). Then, Morpheus core system translates this AST into DataFrame operations with schema and data-type handling. Morpheus provides a graph data definition language (GDDL) for handling schema mapping. Particularly, GDDL expresses property graph types and maps between those types and the relational data sources. Moreover, the Morpheus core system manages importing graph data that can reside in different Spark storage

---

[16]https://github.com/graphframes/graphframes.

[17]https://github.com/opencypher/morpheus.

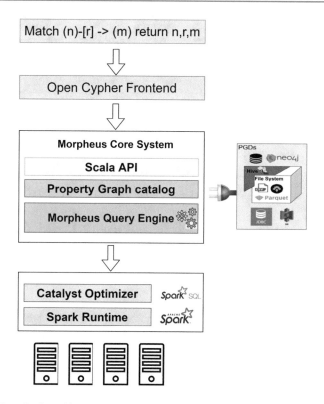

Fig. 4.16 Morephus's architecture

back-ends such as HDFS, Hive, relational databases using JDBC, or Neo4j (i.e., Morpheus property graph data sources PGDs), and exporting these property graphs directly back to those Spark storage back-ends. The native Spark catalyst optimizer is used in Morpheus pipeline for making various core query optimizations for the generated relational plan of operations. Morpheus runs these optimized query plans on the Spark cluster using distributed Spark runtime environment.

4.6   Gradoop

Gradoop[18] is a distributed graph querying and processing framework which has been initially based on the Hadoop framework, then it has been adopted based on the Apache Flink big data processing framework [95]. It is designed to support the extended property graph model (EPGM). The main programming abstractions supported by Gradoop are *LogicalGraph* and *GraphCollection*. In addition, it supports a set of graph operators to express transformations among them [96].

---

[18]https://github.com/dbs-leipzig/gradoop.

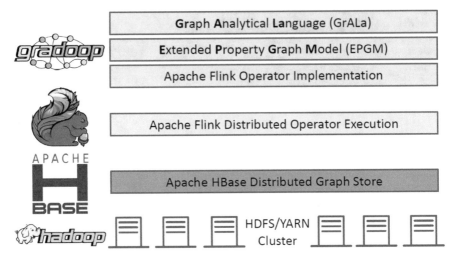

Fig. 4.17 Gradoop's architecture

Gradoop supports the `Cypher` declarative graph query language [97] where its query engine is fully integrated so that pattern matching can be used in combination with other analytical graph operators provided by the framework [98, 99]. In particular, analytical programs are defined within using Gradoop's Graph Analytical Language (GrALa) which supports operators for accessing graphs in the underlying storage as well as for applying graph operations and analytical graph algorithms to them. Operator execution is carried out by the distributed execution engine which spreads the computation across the available machines. When the computation of an analytical program is completed, results may be either written back to the storage layer or presented to the user (Fig. 4.17).

In general, Flink uses a dataflow optimizer that can choose between various join strategies (partitioning vs. broadcast, hash-join vs. sort-merge-join) and reuses data partitions to minimize data shuffling [52]. Gradoop extends this optimizer by providing graph statistics that are utilized for efficient operator reordering optimization, which is crucial for query performance. In particular, the Gradoop's query planner follows a greedy approach by decomposing the query into sets of vertices and edges and constructing a bushy query plan by iteratively joining embeddings and choosing the query plan that minimizes the size of intermediate results. Vertices and edges that are covered by that plan are removed from the initial sets until there is only one plan left. Gradoop provides several implementations for graph storage including file-based storage (e.g., CSV and JSON), HDFS and HBase.

4.7   Other Systems

Other new systems that are not based on Hadoop, Pregel, and GraphLab have been introduced to address different use cases. For example, *Trinity*[19] is a memory-based distributed system which focuses on optimizing communication and memory cost under the assumption that the whole graph is partitioned across a memory cloud [88]. Trinity is designed to support fast graph exploration as well as efficient parallel graph computations. In particular, Trinity organizes the memory of multiple machines into a globally addressable, distributed memory address space (a memory cloud) to support large graphs. In addition, Trinity leverages graph access patterns in both online and offline computation to optimize memory and communication for best performance. A Trinity system consists of slaves, proxies, and clients. A Trinity slave stores graph data and performs computation on the data. Specifically, each slave stores a portion of the data and processes messages received from other slaves, proxies, or clients. A Trinity proxy only handles messages but does not own any data. It usually serves as a middle tier between slaves and clients. A Trinity client is responsible for enabling users to interact with the Trinity cluster. It is a user interface tier between the Trinity system and end users. The memory cloud is essentially a distributed key-value store which is supported by a memory storage module and a message passing framework. Trinity supports a language called TSL (Trinity Specification Language) that bridges the graph model and the data storage. Due to the diversity of graphs and the diversity of graph applications, it is usually hard to support efficient general-purpose graph computation using a fixed graph schema. Therefore, instead of using a fixed graph schema with fixed computation models, Trinity let users define the graph schema, communication protocols, and computation paradigms through TSL.

Signal/Collect[20] is a vertex-centric programming model where graph algorithms are decomposed into two operations on a vertex: (1) Signaling along edges to inform neighbors about changes in vertex state and (2) Collecting the received signals to update the vertex state [90]. In the Signal/Collect programming model, all computations are executed on a compute graph, where the vertices are the computational units that interact by the means of signals that flow along the edges. Vertices collect the signals and perform some computation on them and then signal their neighbors in the compute graph. Signal/Collect supports both synchronous and asynchronous scheduling of the signal and collect operations. It can also both parallelize computations on multiple processor cores and distribute computations over a commodity cluster. Internally, the system uses the *Akka*[21] distributed actor framework for message passing. The scheduling of operations and message passing is done within workers. A vertex stores its outgoing edges, but neither the vertex nor its outgoing edges have access to the target vertices of the edges. In order

---

[19]http://research.microsoft.com/en-us/projects/trinity/.

[20]https://code.google.com/p/signal-collect/.

[21]http://akka.io/.

to efficiently support parallel and distributed executions, modifications to target vertices from the model are translated to messages that are passed via a message bus. Every worker and the coordinator have one pluggable message bus each that takes care of sending signals and translating graph modifications to messages. At the execution time, the graph is partitioned by using a hash function on the vertex IDs. In addition, in order to balance the partitions, Signal/Collect employs the optimizations introduced by GPS [84]. The default storage implementation of Signal/Collect keeps the vertices in memory for fast read and write access. In principle, graph loading can be done sequentially from a coordinator actor or also in parallel, where multiple workers load different parts of the graph at the same time. In addition, specific partitions can be assigned to be loaded by particular workers so that each worker can load its own partition which increases the locality of the loading process.

TurboGraph[22] is a disk-based graph engine which is designed to process billion-scale graphs very efficiently by using modern hardware on a single PC [100]. Therefore, similar to GraphChi, TurboGraph belongs to the centralized category of systems. In particular, TurboGraph is a parallel graph engine that exploits the full parallelism of multicore and FlashSSD IO in addition to the full overlap of CPU processing and I/O processing. By exploiting multicore CPUs, the system can process multiple CPU jobs at the same time, while by exploiting the FlashSSDs, the system can process multiple I/O requests in parallel by using the underlying multiple flash memory packages. In addition, the system applies a parallel execution model, called *pin-and-slide*, which implements the column view of the matrix–vector multiplication. By interpreting the matrix-vector multiplication in the column view, the system can restrict the computation to just a subset of the vertices, utilizing two types of thread pools, the execution thread pool and the asynchronous I/O callback thread pool along with a buffer manager. Specifically, given a set of vertices, the systems start by identifying the corresponding pages for the vertices and then pin those pages in the buffer pool. By exploiting the buffer manager of the storage engine, some pages that were read before can exist in the buffer pool, and the system can guarantee that those pages pinned are to be resident in memory until they are explicitly unpinned. The system then issues parallel asynchronous I/Os to the FlashSSD for pages which are not in the buffer pool. As soon as the I/O request for each page is completed, a callback thread processes the CPU processing of the page. As soon as either an execution thread or a callback thread finishes the processing of a page, it unpins the page, and an execution thread issues an asynchronous I/O request to the FlashSSD. With this mechanism, the system can slide the processing window one page at a time for all pages corresponding to the input vertices essentially enabling it to fully utilize both CPU and FlashSSD I/O parallelism and fully overlap CPU processing and I/O processing.

The *GRACE* system,[23] implemented in C++, is another centralized system that has been introduced as a general parallel graph processing framework that provides

---

[22]http://wshan.net/turbograph.

[23]http://www.cs.cornell.edu/bigreddata/grace/.

an iterative synchronous programming model for developers. GRACE follows batch-style graph programming frameworks to insulate users from low-level details by providing a high-level representation for graph data and letting users specify an application as a set of individual vertex update procedures. The programming model captures data dependencies using messages passed between neighboring vertices like the BSP model. GRACE combines synchronous programming with asynchronous execution for large-scale graph processing by separating application logic from execution policies.

Blogel[24] has been introduced as a block-centric graph-parallel abstraction [101]. In principle, Blogel is conceptually following the same model of Pregel but works in coarser-grained graph units called blocks where a block refers to a connected subgraph of the graph, and message exchanges occur among blocks. Blogel is designed to address the problem of skewed degree distribution and to solve the heavy communication problem caused by high density, since the neighbors of many vertices are now within the same block, and hence they do not need to send/receive messages to/from each other. In addition, Blogel is able to effectively handle large-diameter graphs, since messages now propagate in the much larger unit of blocks instead of single vertices, and thus the number of rounds is significantly reduced. Also, since the number of blocks is usually orders of magnitude smaller than the number of vertices, the workload of a worker is significantly less than that of a vertex-centric algorithm.

4.8 Large-Scale RDF Processing Systems

The Resource Description Framework (RDF) [103] is a graph-based data model which is gaining widespread acceptance in various domains such as semantic web, knowledge management, bioinformatics, business intelligence, and social networks. For example, RDF-based knowledge bases with millions and billions of facts from DBpedia,[25] Probase,[26] and YAGo[27] are now publicly available. In principle, RDF has been designed as a flexible model for schema-free information; data items are represented as triples in the form (S, P, O), where S stands for subject, P for predicate, and O for object. In practice, a collection of triples can be modeled as a directed graph, with vertices denoting subjects and objects, and edges representing predicates (Fig. 4.18).

The SPARQL query language [104] is the official W3C standard for querying and extracting information from RDF graphs [104]. It represents the counterpart to *select-project-join* queries in the relational model. It is based on a powerful *graph matching* facility that allows the binding of variables to components in the

---

[24]http://www.cse.cuhk.edu.hk/blogel/.

[25]http://wiki.dbpedia.org/.

[26]http://research.microsoft.com/en-us/projects/probase/.

[27]https://datahub.io/dataset/yago.

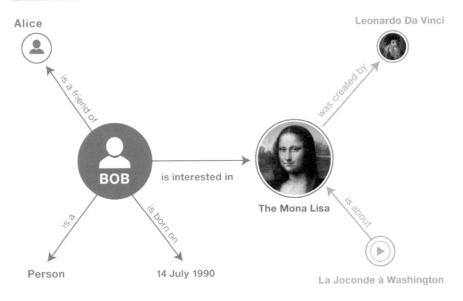

Fig. 4.18 An example of a graph of triples [102]

input RDF graph. Accordingly, satisfying a SPARQL query is usually deemed as a subgraph pattern matching problem. With the growing sizes of RDF databases, several approaches have been introducing for querying large-scale RDG graph databases using big data processing systems [105, 106].

4.8.1   NoSQL-Based RDF Systems

Several approaches have been exploiting NoSQL database systems for building scalable RDF management systems. Figure 4.19 gives an overview of RDF systems classified according to their underlying NoSQL database design. For example, JenaHBase [107] uses HBase, a NoSQL column family store, to provide various custom-built RDF data storage layouts which cover various trade-offs in terms of query performance and physical storage (Fig. 4.20). In particular, JenaHBase designs several HBase tables with different schemas to store RDF triples. The simple layout uses three tables each indexed by subjects, predicates, and objects. For every unique predicate, the vertically partitioned layout creates two tables where each of them is indexed by subjects and objects. The indexed layout uses six tables representing the six possible combinations of indexing RDF triples. The hybrid layout combines both the simple and vertical partitioning layouts. The hash layout combines the hybrid layout with hash values for nodes and a separate table maintaining hash-to-node encodings. For each of these layouts, JenaHBase processes all operations (e.g., loading triples, deleting triples, querying) on an

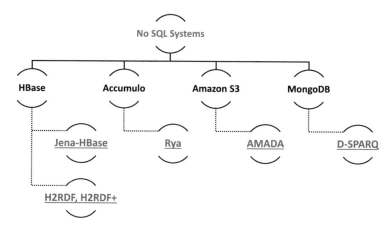

Fig. 4.19 NoSQL-based RDF systems

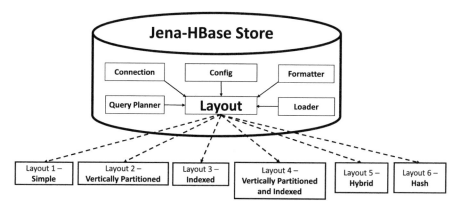

Fig. 4.20 The architecture of JenaHBase system [107]

RDF graph by implicitly converting them into operations on the underlying storage layout.

The Rya system [108] has been built on top of Accumulo, a distributed key-value and column-oriented NoSQL store which supports the ordering of keys in a lexicographical ascending order. Accumulo orders and partitions all key-value pairs according to the row ID part of the key. Rows with similar IDs are grouped into the same node for efficient and faster access. Rya stores the RDF triple (subject, predicate, and object) in the Row ID part of the Accumulo tables. In addition, it indexes the triples across three separate tables (SPO, POS, and OSP) that support all the permutations of the triple pattern. These tables store the triple in the Accumulo Row ID and order the subject, predicate, and object differently for each table. This approach exploits the row-sorting mechanism of Accumulo to efficiently store and query triples across multiple Accumulo tables. SPARQL queries are evaluated using indexed nested loops join operations.

AMADA [109] has been presented as a platform for RDF data management which is implemented on top of the Amazon Web Services (AWS) cloud platform. It is designed as a Software-as-a-Service (SaaS) which allows users to upload, index, store, and query RDF data. In particular, RDF data is stored using Amazon Simple Storage Service (S3).[28] The S3 interface assigns an URL to each dataset which can be later used during the query processing on EC2 nodes. To synchronize the distributed query processing AMADA uses Amazon Simple Queue Service (SQS) providing an asynchronous message-based communication. AMADA builds its own data indexes using SimpleDB, a simple database system supporting SQL-style queries based on a key-value model that supports single-relation queries, i.e., without joins. In AMADA, the query execution is performed using virtual machines within the Amazon Elastic Compute Cloud (EC2). In practice, once a query is submitted to the system, it is sent to a query processor module running on an EC2 instance, which performs a lookup to the indexes in SimpleDB in order to find out the relevant indexes for answering the query, and evaluates the query against them. Results are written in a file stored in S3, whose URI is sent back to the user to retrieve the query answers.

CumulusRDF [110] is an RDF store which provides triple pattern lookups, a linked data server and proxy capabilities, bulk loading, and querying via SPARQL. The storage back-end of CumulusRDF is Apache Cassandra, a NoSQL database management system originally developed by Facebook [111]. Cassandra provides decentralized data storage and failure tolerance based on replication and failover. Cassandra's data model consists of nestable distributed hash tables. Each hash in the table is the hashed key of a row and every node in a Cassandra cluster is responsible for the storage of rows in a particular range of hash keys. The data model provides two more features used by CumulusRDF: super columns, which act as a layer between row keys and column keys, and secondary indices that provide value-key mappings for columns. The index schema of CumulusRDF consists of four indices (SPO, PSO, OSP, CSPO) to support a complete index on triples and lookups on named graphs (contexts). The indices provide fast lockup for all variants of RDF triple patterns. The indices are stored in a "flat layout" utilizing the standard key-value model of Cassandra. CumulusRDF does not use dictionaries to map RDF terms but instead stores the original data as column keys and values. Thereby, each index provides a hash-based lookup of the row key, a sorted lookup on column keys and values, thus enabling prefix lookups. CumulusRDF uses the Sesame query processor[29] to provide SPARQL query functionality. A stock Sesame query processor translates SPARQL queries to index lookups on the distributed Cassandra indices; Sesame processes joins and filter operations on a dedicated query node.

D-SPARQ [112] has been presented as a distributed RDF query engine on top of MongoDB, a NoSQL document database.[30] D-SPARQ constructs a graph from

---

[28]https://aws.amazon.com/s3.

[29]http://www.openrdf.org/.

[30]https://www.mongodb.com/.

the input RDF triples, which is then partitioned, using hash partitioning, across the machines in the cluster. After partitioning, all the triples whose subject matches a vertex are placed in the same partition as the vertex. In other words, D-SPARQ uses hash partitioning based on subject. In addition, similar to [113], a partial data replication is then applied where some of the triples are replicated across different partitions to enable the parallelization of the query execution. Grouping the triples with the same subject enables D-SPARQ to efficiently retrieve triples which satisfy subject-based star patterns in one read call for a single document. D-SPARQ also uses indexes involving subject-predicate and predicate-object. The selectivity of each triple pattern plays an important role in reducing the query runtime during query execution by reordering the individual triple patterns within a star pattern. Thus, for each predicate, D-SPARQ keeps a count of the number of triples involving that particular predicate.

H2RDF+ [114] introduced an indexing scheme over HBase,[31] NoSQL Key-Value store, to expedite SPARQL query processing. H2RDF+ maintains eight indexes, each stored as a table in HBase. Six out of the eight indexes represent all permutations of RDF (S, P, O) elements. The other two indices maintain aggregated index statistics, which are exploited to estimate selectivity of triple patterns and output sizes and costs of joins. H2RDF+ partitions RDF datasets using the HBase internal partitioner, but does not partition queries. In particular, a query is sent as is to each machine, yet gets optimized within each cluster machine. H2RDF+ shuffles intermediate data and utilizes a Hadoop-based sort-merge join algorithm to join them. Lastly, H2RDF+ runs simple queries over single machines and complex ones over a fixed number of distributed machines.

4.8.2   Hadoop-Based RDF Systems

Several systems have been exploiting the Hadoop framework for building scalable RDF processing engines. For example, HadoopDB-RDF [115, 116] partitions the RDF graph across number of nodes where each node is running a centralized RDF query processor, RDF-3X [117]. To minimize intermediate data shuffling during the query processing, HadoopDB-RDF uses graph partitioning by vertex instead of simple hash partitioning by subject, object, or predicate. As a result, vertices that are nearby in the RDF graph can be naturally collocated (if included in the same partition) and mapped onto the same machine. Subsequently, queries that only involve paths within the bounds of partitions can be executed in an embarrassingly parallel fashion. In addition, triples at the boundary of each partition are replicated according to a mechanism denoted as *n-hop* guarantee. HadoopDB-RDF leverages on the HadoopDB execution engine [69] to handle the splitting of query execution across the high performance single node database systems and the Hadoop data processing framework (Fig. 4.21).

---

[31]https://hbase.apache.org/.

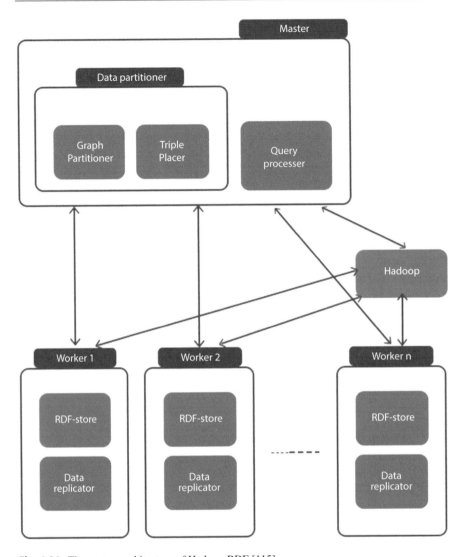

Fig. 4.21 The system architecture of Hadoop-RDF [115]

CliqueSquare [118, 119] is another Hadoop-based RDF data management platform for storing and processing big RDF datasets. With the central goal of minimizing the number of MapReduce jobs and the data transfer between nodes during query evaluation, CliqueSquare exploits the built-in data replication mechanism of the Hadoop Distributed File System (HDFS). Each of its partition has three replicas by default, to partition the RDF dataset in different ways. In particular, for the first replica, CliqueSquare partitions triples based on their subject, property, and object values. For the second replica, CliqueSquare stores all subject, property, and object partitions of the same value within the same node. Finally, for the third replica, CliqueSquare groups all the subject partitions within a node by the value of the property in their triples. Similarly, it groups all object partitions based on their property values. In addition, CliqueSquare implements a special treatment for triples whose property is *rdf:type*, by translating them into an unwieldy large property partition. CliqueSquare then splits the property partition of *rdf:type* into several smaller partitions according to their object value. For SPARQL query processing, CliqueSquare relies on a clique-based algorithm which produces query plans that minimize the number of MapReduce stages. The algorithm is based on the variable graph of a query and its decomposition into clique subgraphs. The algorithm works in an iterative way to identify cliques and to collapse them by evaluating the joins on the common variables of each clique. The process ends when the variable graph consists of only one node. Since triples related to a particular resource are co-located on one node CliqueSquare can perform all first-level joins in RDF queries (SS, SP, SO, PP, PS, PO, etc.) locally on each node and reduce the data transfer through the network. In particular, it allows queries composed of 1-hop graph patterns to be processed in a single MapReduce job, which enables a significant performance competitive advantage.

The PigSPARQL system [120] compiles SPARQL queries into the Pig query language [121], a data analysis platform over the Hadoop framework. Pig uses a fully nested data model and provides relational style operators (e.g., filters and joins). In PigSPARQL, a SPARQL query is parsed to generate an abstract syntax tree which is subsequently compiled into a SPARQL algebra tree. Using this tree, PigSPARQL applies various optimizations on the algebra level such as the early evaluation of filters and using the selectivity information for reordering the triple patterns. Finally, PigSPARQL traverses the optimized algebra tree bottom-up and generates an equivalent sequence of Pig Latin expressions for every SPARQL algebra operator. For query execution, Pig automatically maps the resulting Pig Latin script onto a sequence of Hadoop jobs. An advantage of taking PigSPARQL as an intermediate layer that uses Pig between SPARQL and Hadoop is being independent of the actual Hadoop version or implementation details. RAPID+ [122] is another Pig-based system that uses an algebraic approach for optimizing and evaluating SPARQL queries on top of the Hadoop framework. It uses a data model and algebra (the Nested TripleGroup Data Model and Algebra (NTGA) [123]) which includes support for expressing graph pattern matching queries.

4.8.3   Spark-Based RDF Systems

Several systems have been designed to exploit the Spark framework for building scalable RDF processing engines. For example, S2RDF[32] (*SPARQL* on *Spark* for *RDF*) [124] introduced a relational partitioning schema for encoding RDF

---

[32]http://dbis.informatik.uni-freiburg.de/S2RDF.

data called ExtVP (the *Ext*ended *V*ertical *P*artitioning) that extends the Vertical
Partitioning (VP) schema introduced by Abadi et al. [125] and uses a semi-join
based preprocessing to efficiently minimize query input size by taking into account
the possible join correlations between the underlying encoding tables of the RDF
data and join indices [126]. In particular, ExtVP precomputes the possible join
relations between partitions (i.e., tables). The main goal of ExtVP is to reduce
the unnecessary I/O operations, comparisons, and memory consumption during
executing join operations by avoiding the dangling tuples in the input tables of the
join operations, i.e., tuples that do not find a join partner. In particular, S2RDF
determines the subsets of a *VP* table VP_{p1} that are guaranteed to find at least
one match when joined with another VP table VP_{p2}, where $p1$ and $p2$ are
query predicates. S2RDF uses this information to precompute a number of semi-
join reductions [127] of VP_{p1} . The relevant semi-joins between tables in VP
are determined by the possible joins that can occur when combining the results
of triple patterns during query execution. Clearly, ExtVP comes at the cost of
some additional storage overhead in comparison to the basic vertical partitioning
techniques. Therefore, ExtVP does not use exhaustive precomputations for all the
possible join operations. Instead, an optional selectivity threshold for ExtVP can
be specified to materialize only the tables where reduction of the original tables
is large enough. This mechanism facilitates the ability to control and reduce the
size overhead while preserving most of its performance benefit. S2RDF uses the
Parquet[33] columnar storage format for storing the RDF data on the Hadoop
Distributed File System (HDFS). S2RDF is built on top of Spark. The query
evaluation of S2RDF is based on SparkSQL [49], the relational interface of Spark.
It parses a SPARQL query into the corresponding algebra tree and applies some
basic algebraic optimizations (e.g., filter pushing) and traverses the algebraic tree
bottom-up to generate the equivalent Spark SQL expressions. For the generated
SQL expression, S2RDF can use the precomputed semi-join tables, if they exist,
or alternatively uses the base encoding tables.

SparkRDF [128, 129] is another Spark-based RDF engine which partitions the
RDF graph into MESGs (Multi-layer Elastic SubGraphs) according to relations (R)
and classes (C) by building five kinds of indices (C,R,CR,RC,CRC) with different
granularities to support efficient evaluation for the different query triple patterns.
SparkRDF creates an index file for every index structure and stores such files
directly in the HDFS, which are the only representation of triples used for the query
execution. These indices are modeled as RDSGs (Resilient Discreted SubGraphs),
a collection of in-memory subgraphs partitioned across nodes. SPARQL queries
are evaluated over these indices using a series of basic operators (e.g., filter,
join). All intermediate results are represented as RDSGs and maintained in the
distributed memory to support faster join operations. SparkRDF uses a selectivity-
based greedy algorithm to build a query plan with an optimal execution order of
query triple patterns that aims to effectively reduce the size of the intermediate

---

[33]https://parquet.apache.org/.

results. In addition, it uses a location-free pre-partitioning strategy that avoids the expensive shuffling cost for the distributed join operations. In particular, it ignores the partitioning information of the indices while repartitioning the data with the same join key to the same node.

The S2X (SPARQL on Spark with GraphX) [130] RDF engine has been implemented on top of GraphX [51], an abstraction for graph-parallel computation that has been augmented to Spark [44]. It combines graph-parallel abstractions of GraphX to implement the graph pattern matching constructs of SPARQL. A similar approach has been followed by Goodman and Grunwald [131] for implementing an RDF engine on top the GraphLab framework, another graph-parallel computation platform [87]. Naacke et al. [132] compared five Spark-based SPARQL query processing based on different join execution models. The results showed that hybrid query plans combining partitioned join and broadcast joins improve query performance in almost all cases.

4.8.4   Other Distributed RDF Systems

Trinity.RDF [133] has been introduced as a system for storing RDF datasets in their native graph form on top of Trinity [88], a distributed in-memory key-value store. In particular, Trinity.RDF models RDF data as an in-memory graph and replaces joins with graph explorations in order to prune the search space and avert generating unnecessary intermediate results. It decomposes a given SPARQL query, Q, into a sequence of patterns so that the bindings of a current pattern can exploit the bindings of a previous pattern (i.e., patterns are not executed independently). Patterns are ordered using a query optimizer so as to reduce the volume of bindings.

TripleRush [134, 135] is based on the graph processing framework Signal/Collect [136], a parallel graph processing system written in Scala. TripleRush evaluates queries by routing partially matched copies of the query through an index graph. By routing query descriptions to data, the system eliminates joins in the traditional sense. TripleRush implements three kinds of Signal/Collect vertices: (1) *Triple Vertices* represent RDF triples; each vertex contains a subject, predicate, and object. (2) *Index Vertices* for triple patterns that route to triples. Each vertex contains a triple pattern (with one or more positions as wildcards); these vertices build a graph from a match-it-all triple pattern to actual tipples. (3) *Query Vertices* to coordinate the query execution process. Such vertices are created for each query executed in the system. The vertex then initiates a query traversal process through the index graph before returning the results. The most distinguished feature of TripleRush is the ability to inherently divide a query among many processing units.

AdHash [137, 138] is another distributed in-memory RDF engine which initially applies lightweight hash partitioning that distributes triples of the RDF triples by hashing on their subjects. AdHash attempts to improve the query execution times by increasing the number of join operations that can be executed in parallel without data communication through utilizing hash-based locality. In particular, the join patterns on subjects included in a query can be processed in parallel. The locality-aware

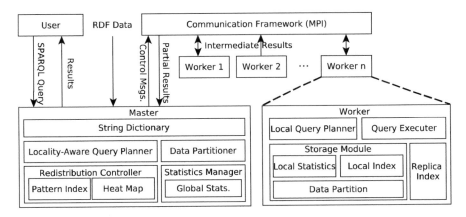

Fig. 4.22 The architecture of AdHash system [137]

query optimizer exploits this property to build a query evaluation plan that reduces the size of intermediate results transferred among the worker nodes. In addition, AdHash continuously monitors the data access patterns of the executed workload and dynamically adapts to the query workload by incrementally redistributing and replicating the frequently used partitions of the graphs. The main goal for the adaptive dynamic strategy of AdHash is to effectively minimize or eliminate the data communication cost for future queries. Therefore, hot patterns are redistributed and potentially replicated to allow future workloads which contain them to be evaluated in parallel by all worker nodes without any data transfer. To efficiently manage the replication process, AdHash specifies a budget constraint and uses an eviction policy for the redistributed patterns. As a result, AdHash attempts to overcome the disadvantages of static partitioning schemes and dynamically reacts with changing workloads. Figure 4.22 illustrates the architecture of the AdHash RDF engine. In this architecture, the master starts by partitioning the data across the worker nodes and gathering global statistical information. In addition, the master node is responsible for receiving queries from users, generating execution plans, coordinating worker nodes, collecting final results, and returning the results to users. The statistics manager maintains statistics about the RDF graph that are exploited during the global query planning and adaptive repartitioning purposes. These statistics are distributedly gathered during the bootstrapping phase. The redistribution controller monitors the executed query workload in the form of heat maps and starts an adaptive Incremental ReDistribution (IRD) process for hot patterns. In this process, only data which is retrieved by the hot patterns are redistributed and potentially replicated across the worker nodes [137]. In principle, a redistributed hot pattern can be answered by all workers in parallel without communication. The locality-aware query planner uses the global statistics and the pattern index from the redistribution controller to decide if a query, in whole or partially, can be processed without communication. Queries that can be fully answered without communication are planned and executed by each worker independently. On the other hand, for queries that require communication, the planner exploits the hash-based data locality and the query structure to find a plan that minimizes communication and the number of distributed joins [137].

The TriAD (Triple-Asynchronous-Distributed) system [139] uses a main-memory shared-nothing architecture and is based on an asynchronous Message Passing protocol. TriAD applies a classical master-slave architecture in which the slave nodes autonomously and asynchronously exchange messages among them to evaluate multiple join operations in parallel. Relying on asynchronous communication allows the sibling execution paths of a query plan to be processed in a freely multi-threaded fashion and only get merged (i.e., get synchronized) when the intermediate results of entire execution paths are joined. TriAD employs six comprehensive combinations of indexing over the RDF elements. These indices are maintained into a distributed main-memory data structure where each index is first hash-partitioned according to its join key and then locally sorted in lexicographic order. Therefore, TriAD can perform efficient, distributed merge-joins over the hash-partitioned permutation lists. In addition, TriAD uses join-ahead pruning using an additional RDF summary graph which is deployed at the master node, in order to prune entire partitions of triples from the SPO lists that cannot contribute to the results of a given SPARQL query. TriAD uses a bottom-up dynamic programming mechanism for join-order enumeration and considers the locality of the index structures at the slave nodes, the data exchange cost of the intermediate results, and the option to execute sibling paths of the query plan in a multi-threaded fashion, in order to estimate the query execution plan with the cheapest cost.

Partout [140] is a distributed engine that relies on a workload-aware partitioning strategy for RDF data by allowing queries to be executed over a minimum number of machines. Partout exploits a representative query load to collect information about frequently co-occurring subqueries and for achieving optimized data partitioning and allocation of the data to multiple nodes. The architecture of Partout consists of a coordinator node and a cluster of n hosts that store the actual data. The coordinator node is responsible for distributing the RDF data among the host nodes, designing an efficient distributed query plan for a SPARQL query, and initiating query evaluation. The coordinator does not have direct access to the actual data but instead utilizes global statistics of the RDF data, generated at partitioning time, for query planning. Each of the host nodes runs a triple store, RDF-3X [141]. Queries are issued at the coordinator, which is responsible for generating a suitable query plan for distributed query execution. The data is located at the hosts which are hosting the data partitions. Each host executes part of the query over its local data and sends the results to the coordinator, which will finally hold the query result. Partout's global query optimization algorithm avoids the need for a two-step approach by starting with a plan optimized with respect to the selectivities of the query predicates and then applying heuristics to obtain an efficient plan for the distributed setup. Each host relies on the RDF-3X optimizer for optimizing its local query plan.

The DREAM (Distributed RDF Engine with Adaptive Query Planner and Minimal Communication) system [142, 143] has been designed with the aim of avoiding partitioning RDF graphs and partitions SPARQL queries only, thus attempting to combine the advantages of the centralized and distributed RDF systems. DREAM stores a complete dataset at each cluster machine and employs a query planner that effectively partitions any SPARQL query, Q. In particular, DREAM partitions SPARQL queries rather than partitioning RDF datasets. This is achieved by using rule- and cost-based query planner that uses statistical information of the RDF database. Specifically, the query planner transforms Q into a graph, G, decomposes G into sets of subgraphs, each with a basic two-level tree structure, and maps each set to a separate machine. Afterwards, all machines process their sets of subgraphs in parallel and coordinate with each other to return the final result. No intermediate data is shuffled whatsoever and only minimal control messages and metadata[34] are exchanged. To decide upon the number of sets (which dictates the number of machines) and their constituent subgraphs (i.e., G's *graph plan*), the query planner enumerates various possibilities and selects a plan that will expectedly result in the lowest network and disk costs for G. This is achieved through utilizing a cost model that relies on RDF graph statistics. Using the above approach, DREAM is able to select different numbers of machines for different query types, hence, rendering it adaptive.

Cheng and Kotoulas [144] presented a hybrid method for processing RDF that combines similar-size and graph-based partitioning strategies. With similar-size partitioning, it places similar volumes of raw triples on each computation node without a global index. In addition, graph partitioning algorithms are used to partition RDF data in a manner such that triples close to each other can be assigned to the same computation node. In practice, the main advantage of similar-size partitioning is that it allows for fast loading data, while graph-based partitioning allows to achieve efficient query processing. A two-tier index architecture is adopted. The first tier is a lightweight *primary index* which is used to maintain low loading times. The second tier is a series of dynamic, multi-level secondary indexes, evaluated during query execution, which is utilized for decreasing or removing the inter-machine data transfer for subsequent operations which maintain similar graph patterns. Additionally, this approach relies on a set of parallel mechanisms which combine the loading speed of similar-size partitioning with the execution speed of graph-based partitioning. For example, it uses fixed-length integer encoding for RDF terms and indexes that are based on hash tables to increase access speed. The indexing process does not use network communication in order to increase the loading speed. The local lightweight primary index is used to support very fast retrieval and avoid costly scans, while the secondary indexes are used to support nontrivial access patterns that are built dynamically, as a byproduct of query execution, to amortize costs for common access patterns.

The DiploCloud system [145] has been designed to use a hybrid storage structure by co-locating semantically related data to minimize inter-node operations. The co-

---

[34]DREAM uses RDF-3X [141] at each slave machine and communicates only triple ids (i.e., metadata) across machines. Locating triples using triple ids in RDF-3X is a straightforward process.

located data patterns are mined from both instance and schema levels. DiploCloud uses three main data structures: molecule clusters, template lists, and a molecule index. *Molecule clusters* extend property tables to form RDF subgraphs that group sets of related URIs in nested hash tables and to co-locate data corresponding to a given resource. *Template lists* are used to store literals in lists, like in a columnar database system. Template lists allow to process long lists of literals efficiently; therefore, they are employed mainly for analytics and aggregate queries. The molecule index serves to index URIs based on the molecule cluster to which they belong. In the architecture of DiploCloud, the Master node is composed of three main subcomponents: a key index encoding URIs into IDs, a partition manager, and a distributed query executor. The Worker nodes of the system hold the partitioned data and its corresponding local indices. The workers store three main data structures: a type index (grouping keys based on their types), local molecule clusters, and a molecule index. The worker nodes run subqueries and send results to the Master node. The data partitioner of DiploCloud relies on three molecule-based data partitioning techniques: (1) *Scope-k Molecules* manually define the size for all molecules, (2) *Manual Partitioning* where the system takes an input manually defined shapes of molecules, (3) *Adaptive Partitioning* starts with a default shape of molecules and adaptively increases or decreases the size of molecules based on the workload. Queries that are composed of one Basic Graph Pattern (e.g., star-like queries) are executed in parallel without any central coordination. For queries requiring distributed joins, DiploCloud picks one of the two executions strategies: (1) if the intermediate result set is small, DiploCloud ships everything to the Master node that performs the join and (2) if the intermediate result set is large, DiploCloud performs a distributed hash-join.

Blazegraph[35] is an open-source triplestore written in Java which is designed to scale horizontally by distributing data with dynamic key-range partitions. It also supports transactions with multiversion concurrency control relying on timestamps to detect conflicts. It maintains three RDF indices (SPA, POS, OSP) and leverages a B+Tree implementation. Those indices are dynamically partitioned into key-range shards that can be distributed between nodes in a cluster. Its scale-out architecture is based on multiple services. The shard locator service maps each key-range partition to a metadata record that allows to locate the partition. A transaction service coordinates locks to provide isolation. A client service, finally, allows to execute distributed tasks. The query execution process starts with the translation of a SPARQL query to an abstract syntax tree (AST). Then, the tree is rewritten to optimize the execution. Finally, it is translated to a physical query plan, vectorized, and submitted for execution.

---

[35]https://www.blazegraph.com.

Large-Scale Stream Processing Systems

5.1   The Big Data Streaming Problem

In practice, as the world gets more instrumented and connected, we are witnessing a flood of digital data that is getting generated from several hardware (e.g., sensors) or software in the format of flowing stream of data. Examples of these phenomena are crucial for several applications and domains including financial markets, surveillance systems, manufacturing, smart cities, and scalable monitoring infrastructure. In all of these applications and domains, there is a crucial requirement to collect, process, and analyze big streams of data in order to extract valuable information, discover new insights in real-time, and to detect emerging patterns and outliers [146]. For example, in 2017, Walmart, the world's biggest retailer with over 20,000 stores in 28 countries reported that it processes more than one million customer transaction every hour, generated by over 200 streams of internal and external data sources, which adds more than 2.5 Petabytes to its databases.[1] The New York Stock Exchange (NYSE), the world's largest stock exchange, has reported trading more than 3.6 billion shares, in more than 2 million trades, on a typical day in November of 2017.[2] According to IDTechEx , by the end of 2019, there were 20 billion sold tags versus 17.5 billion in 2018. IBM has also reported that T-Mobile, a telecommunication company, processed more than 17 billion events including text messages, phone calls, and data traffic over the communication network. Furthermore, the avionics of modern commercial jets outputs 10 terabytes (TB) of data, per engine, during every half-hour of operation. British petroleum (BP) made $7bn annual savings since 2014 by investing in Big Data technologies.[3]

---

[1] https://www.forbes.com/sites/bernardmarr/2017/01/23/really-big-data-at-walmart-real-time-insights-from-their-40-petabyte-data-cloud/.

[2] https://dashboard.theice.com/nyxdata-landing.html.

[3] https://blog.scottlogic.com/2017/07/18/bp-big-data.html.

© The Editor(s) (if applicable) and The Author(s), under exclusive
licence to Springer Nature Switzerland AG 2020
S. Sakr, *Big Data 2.0 Processing Systems*,
https://doi.org/10.1007/978-3-030-44187-6_5

Management and maintenance representing a significant operational cost. They continuously monitoring key operating characteristics of assets and applying advanced algorithms to maximize asset performance and minimize outages. BP deployed extremely large number of sensors to more than 99% of their oil and gas wells. Processing of this data on time helps in scheduling preventative maintenance and delivering replacements parts or any other actions to be taken at just the right time. All of these are just examples that illustrate the data processing challenges that some applications and organizations can face today. In all of these applications, data must be ingested, processed, and analyzed quickly and ideally continuously as the data is continuously produced.

In practice, Internet Of Things (IoT) is considered as the killer applications of big streaming data. IoT is described as network of devices, connect directly with each other to capture, share, and monitor vital data automatically through the cloud [147]. It enables communication between devices, people, and processes to exchange useful information and knowledge that create value for humans. Thus, it is considered as a global network infrastructure linking physical and virtual objects (Fig. 5.1). The world is witnessing a continuous increase in the number of connected

Fig. 5.1 Internet of Things

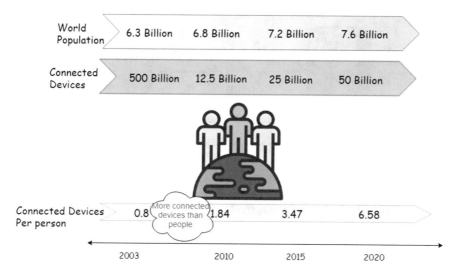

Fig. 5.2 Number of IoT devices

IoT devices (Fig. 5.2) which is expected to reach 75 billion by 2025.[4] In addition, Internet Of Things based applications are adopted on several application domains and thus implementing efficient and effective analytics on the generated massive amounts of data is a crucial requirement for realizing the Smart X phenomena (e.g., Smart Home, Smart Hospital, Smart City, etc.).

In principle, stream computing is a new paradigm necessitated by new data-generating scenarios, such as the ubiquity of mobile devices, location services, and sensor pervasiveness [148]. In general, stream processing systems support a large class of applications in which data are generated from multiple sources and are pushed asynchronously to servers which are responsible for processing them. Data streams are conceptually infinite, ever growing set of data items/events. They require continuous processing to tame the high *velocity* of the data. Therefore, stream processing applications are usually deployed as continuous jobs that run from the time of their submission until their cancellation. It processes the data as it is changing and arriving continuously (data in motion). Figure 5.3 illustrates the difference between the computations performed on static data and those performed on streaming data. In particular, in static data computation, questions are asked of static data. In streaming data computation, data is continuously evaluated by static questions.

---

[4]https://www.statista.com/statistics/471264/iot-number-of-connected-devices-worldwide/.

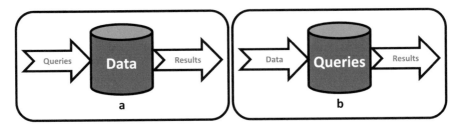

Fig. 5.3 Static data computation versus streaming data computation. (**a**) Static data computation.
(**b**) Streaming data computation

5.2   Hadoop for Big Streams?!

In principle, the fundamental design of the Hadoop framework requires the results
of each single map or reduce task to be *materialized* into a local file before it
can be processed by the following tasks. This materialization step supports the
implementation of a simple and elegant checkpointing/restarting fault tolerance
mechanism. However, it dramatically hurts the performance of applications with
real-time processing requirements. The *MapReduce Online* approach [149, 150]
have been proposed as a modified architecture of the MapReduce framework
in which intermediate data is *pipelined* between operators while preserving the
programming interfaces and fault tolerance models of previous MapReduce frameworks. This pipelining approach provides important advantages to the MapReduce
framework such as:

- The Reducers can begin their processing of the data as soon as it is produced
  by mappers. Therefore, they can generate and refine an approximation of their
  final answer during the course of execution. In addition, they can provide initial
  estimates of the results several orders of magnitude faster than the final results.
- It widens the domain of problems to which MapReduce can be applied. For
  example, it facilitates the ability to design MapReduce jobs that run continuously,
  accepting new data as it arrives and analyzing it immediately (continuous
  queries). This allows MapReduce to be used for applications such as event
  monitoring and stream processing.
- Pipelining delivers data to downstream operators more promptly, which can
  increase opportunities for parallelism, improve utilization, and reduce response
  time.

In this approach, each reduce task contacts every map task upon initiation of the
job and opens a TCP socket which will be used to pipeline the output of the Map
function. As each map output record is produced, the mapper determines to which
partition (reduce task) the record should be sent, and immediately sends it via the
appropriate socket. A reduce task accepts the pipelined data it receives from each

map task and stores it in an in-memory buffer. Once the reduce task learns that every map task has completed, it performs a final merge of all the sorted runs. In addition, the reduce tasks of one job can optionally pipeline their output directly to the map tasks of the next job, sidestepping the need for expensive fault-tolerant storage in HDFS for what amounts to a temporary file. However, the computation of the Reduce function from the previous job and the Map function of the next job cannot be overlapped as the final result of the reduce step cannot be produced until all map tasks have completed, which prevents effective pipelining. Therefore, the reducer treats the output of a pipelined map task as *tentative* until the JobTracker informs the reducer that the map task has committed successfully. The reducer can merge together spill files generated by the same uncommitted mapper, but will not combine those spill files with the output of other map tasks until it has been notified that the map task has committed. Thus, if a map task fails, each reduce task can ignore any tentative spill files produced by the failed map attempt. The JobTracker will take care of scheduling a new map task attempt, as in standard Hadoop. In principle, the main limitation of the *MapReduce Online* approach is that it is based on HDFS. Therefore, it is not suitable for streaming applications, in which data streams have to be processed without any disk involvement. A similar approach has been presented in [151] which defines an *incremental* MapReduce job as one that processes data in large batches of tuples and runs continuously according to a specific window range and slide of increment. In particular, it produces a MapReduce result that includes all data within a window (of time or data size) every slide and considers landmark MapReduce jobs where the trailing edge of the window is fixed and the system incorporates new data into the existing result. Map functions are trivially continuous and process data on a tuple-by-tuple basis. However, before the Reduce function may process the mapped data, the data must be partitioned across the reduce operators and sorted. When the map operator first receives a new key-value pair, it calls the Map function and inserts the result into the latest increment in the map results. The operator then assigns output key-value pairs to reduce tasks, grouping them according to the partition function. Continuous reduce operators participate in the sort as well, grouping values by their keys before calling the Reduce function.

The *Incoop* system [152] has been introduced as a MapReduce implementation that has been adapted for incremental computations which detects the changes on the input datasets and enables the automatic update of the outputs of the MapReduce jobs by employing a fine-grained result reuse mechanism. In particular, it allows MapReduce programs which are not designed for incremental processing to be executed transparently in an incremental manner. To achieve this goal, the design of Incoop introduces new techniques that are incorporated into the Hadoop MapReduce framework. For example, instead of relying on HDFS to store the input to MapReduce jobs, Incoop devises a file system called *Inc-HDFS* (Incremental HDFS) that provides mechanisms to identify similarities in the input data of consecutive job runs. In particular, Inc-HDFS splits the input into chunks whose boundaries depend on the file contents so that small changes to input do not change all chunk boundaries. Therefore, this partitioning mechanism can maximize the opportunities for reusing results from previous computations, while preserving compatibility with HDFS by offering the same interface and semantics. In addition, Incoop controls the granularity of tasks so that large tasks can be divided into smaller subtasks that can be reused even when the large tasks cannot. Therefore, it introduces a new *Contraction phase* that leverages *Combiner* functions to reduce the network traffic by anticipating a small part of the processing done by the Reducer tasks and control their granularity. Furthermore, Incoop improves the effectiveness of memoization by implementing an affinity-based scheduler that applies a work stealing algorithm to minimize the amount of data movement across machines. This modified scheduler strikes a balance between exploiting the locality of previously computed results and executing tasks on any available machine to prevent straggling effects. On the runtime, instances of incremental map tasks take advantage of previously stored results by querying the memoization server. If they find that the result has already been computed, they fetch the result from the location of their memoized output and conclude. Similarly, the results of a reduce task are remembered by storing them persistently and locally where a mapping from a collision-resistant hash of the input to the location of the output is inserted in the memoization server. Since a reduce task receives input from n map tasks, the key stored in the memoization server consists of the hashes of the outputs from all n map tasks that collectively form the input to the reduce task. Therefore, when executing a reduce task, instead of immediately copying the output from the map tasks, the reduce task consults the map tasks for their respective hashes to determine if the reduce task has already been computed in the previous run. If so, that output is directly fetched from the location stored in the memoization server, which avoids the re-execution of that task.

The M^3 system [153] has been proposed to support the answering of continuous queries over streams of data by bypassing the HDFS so that data gets processed only through a main-memory-only data-path and totally avoids any disk access. In this approach, Mappers and Reducers never terminate where there is only one MapReduce job per query operator that is continuously executing. In M^3, query processing is incremental where only the new input is processed, and the change in the query answer is represented by three sets of inserted ($+ve$), deleted ($-ve$), and updated (u) tuples. The query issuer receives as output a stream that represents the deltas (incremental changes) to the answer. Whenever an input tuple is received, it is transformed into a modify operation ($+ve$, $-ve$, or u) that is propagated in the query execution pipeline, producing the corresponding set of modify operations in the answer. Supporting incremental query evaluation requires that some intermediate state be kept at the various operators of the query execution pipeline. Therefore, Mappers and Reducers run continuously without termination, and hence can maintain main-memory state throughout the execution. In contrast to splitting the input data based on its size as it is done when using Hadoop's Input Split functionality, M^3 splits the streamed data based on arrival rates where the Rate Split layer, between the main-memory buffers and the Mappers, is responsible for balancing the stream rates among the Mappers. This layer periodically receives rate statistics from the Mappers and accordingly redistributes the load of processing among Mappers. For instance, a fast stream that can overflow one Mapper should be distributed among two or more Mappers. In contrast, a group of slow streams that would underflow their corresponding Mappers should be combined to feed into only one Mapper. To support fault tolerance, input data is replicated inside the main-memory buffers and an input split is not overwritten until the corresponding Mapper commits. When a Mapper fails, it re-reads its corresponding input split from any of the replica inside the buffers. A Mapper writes its intermediate key-value pairs in its own main memory, and does not overwrite a set of key-value pairs until the corresponding reducer commits. When a Reducer fails, it re-reads its corresponding sets of intermediate key-value pairs from the Mappers.

The *DEDUCE* system [154] has been presented as a middleware that attempts to combine real-time stream processing with the capabilities of a large-scale data analysis framework like MapReduce. In particular, it extends the *IBM's System S* stream processing engine and augments its capabilities with those of the MapReduce framework. In this approach, the input dataset to the MapReduce operator can be either pre-specified at compilation time or could be provided at runtime as a punctuated list of files or directories. Once the input data is available, the MapReduce operator spawns a MapReduce job and produces a list of punctuated list of files or directories, which point to the output data. Therefore, a MapReduce operator can potentially spawn multiple MapReduce jobs over the application's life span but such jobs are spawned only when the preceding job (if any) has completed its execution. Hence, multiple jobs can be cascaded together to create a dataflow of MapReduce operators where the output from the MapReduce operators can be read to provide updates to the stream processing operators.

As a matter of fact, Hadoop is built on top of the Hadoop Distributed File System (HDFS), a distributed file system designed to run on commodity hardware, which is more suitable for batch processing of very large amounts of data rather than for interactive applications. This makes the MapReduce paradigm unsuitable for event-based online Big Data processing architectures, and motivates the need of investigating other different paradigms and novel platforms for large-scale event stream-driven analytics solutions.

5.3   Storm

The *Storm* system has been presented by Twitter as a distributed and fault-tolerant stream processing system that instantiates the fundamental principles of Actor theory [92]. The core abstraction in Storm is the *stream*. A stream is an unbounded sequence of *tuples*. Storm provides the primitives for transforming a stream into a new stream in a distributed and reliable way. The basic primitives that Storm provides for performing stream transformations are *spouts* and *bolts*. A *spout* is a source of streams. A *bolt* consumes any number of input streams, carries out some processing, and possibly emits new streams. Complex stream transformations, such as the computation of a stream of trending topics from a stream of tweets, require multiple steps and thus multiple bolts. A *topology* is a graph of stream

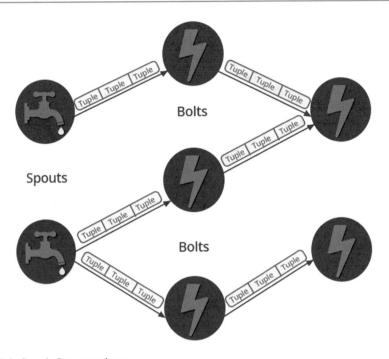

Fig. 5.4 Sample Storm topology

transformations where each node is a spout or bolt. Edges in the graph indicate which bolts are subscribing to which streams. When a spout or bolt emits a tuple to a stream, it sends the tuple to every bolt that has subscribed to that stream. Links between nodes in a topology indicate how tuples should be passed around. Each node in a Storm topology executes in parallel. In any topology, we can specify how much parallelism is required for each node, and then Storm will spawn that number of *threads* across the cluster to perform the execution. Figure 5.4 depicts a sample Storm topology.

The Storm system relies on the notion of *stream grouping* to specify how tuples are sent between processing components. In other words, it defines how that stream should be partitioned among the bolt's tasks. In particular, Storm supports different types of stream groupings such as:

- *Shuffle grouping* where stream tuples are randomly distributed such that each bolt is guaranteed to get an equal number of tuples.
- *Fields grouping* where the tuples are partitioned by the fields specified in the grouping.
- *All grouping* where the stream tuples are replicated across all the bolts.
- *Global grouping* where the entire stream goes to a single bolt.

In addition to the supported built-in stream grouping mechanisms, the Storm system allows its users to define their own custom grouping mechanisms.

Apache Trident[5] has been presented as an extension of Storm which is designed as a high-level abstraction for doing real-time computations. In particular, it plays the same role of high-level batch processing tools such as Pig on the MapReduce ecosystem by providing built-in support for joins, aggregations, grouping, functions, and filters. Storm has influenced the design of several other following systems such as `Apache Heron`[6] and `Alibaba JStorm`.[7]

5.4   Infosphere Streams

Infosphere Streams[8] is a component of the IBM Big Data analytics platform that allows user-developed applications to quickly ingest, analyze, and correlate information as it arrives from thousands of data stream sources. The system is designed to handle up to millions of events or messages per second. It provides a programming model and IDE for defining data sources, and software analytic modules called operators that are fused into processing execution units. It also provides infrastructure to support the composition of scalable stream processing applications from these components. The main platform components are [155]:

- *Runtime environment*: This includes platform services and a scheduler for deploying and monitoring Streams applications across a single host or set of integrated hosts.
- *Programming model*: You can write Streams applications using the Streams Processing Language (SPL), a declarative language. You use the language to state what you want, and the runtime environment accepts the responsibility for determining how best to service the request. In this model, a Streams application is represented as a graph that consists of operators and the streams that connect them.
- *Monitoring tools and administrative interfaces*: Streams applications process data at speeds that are much higher than those that normal operating system monitoring utilities can efficiently handle. InfoSphere Streams provides the tools that can deal with this environment.

SPL,[9] the programming language for InfoSphere Streams is a distributed dataflow composition language. It is an extensible and full-featured language that supports user-defined data types. An InfoSphere Streams continuous application describes a directed graph composed of individual operators that interconnect and operate on multiple data streams. Data streams can come from outside the system

---

[5]https://storm.apache.org/documentation/Trident-API-Overview.html.

[6]https://apache.github.io/incubator-heron/.

[7]http://jstorm.io/.

[8]http://www-03.ibm.com/software/products/en/ibm-streams.

[9]http://www.ibm.com/support/knowledgecenter/SSCRJU_3.2.0/com.ibm.swg.im.infosphere.
streams.spl-language-specification.doc/doc/spl-container.html.

or be produced internally as part of an application. The basic building blocks of
SPL programs are:

- *Stream*: An infinite sequence of structured tuples. It can be consumed by
  operators on a tuple-by-tuple basis or through the definition of a window.
- *Tuple*: A structured list of attributes and their types. Each tuple on a stream has
  the form dictated by its stream type.
- *Stream type*: Specifies the name and data type of each attribute in the tuple.
- *Window*: A finite, sequential group of tuples. It can be based on count, time,
  attribute value, or punctuation marks.
- *Operator*: The fundamental building block of SPL, its operators process data
  from streams and can produce new streams.

Figure 5.5 illustrates the InfoSphere Stream's runtime view of SPL programs.
In this architecture, an operator represents a reusable stream transformer that

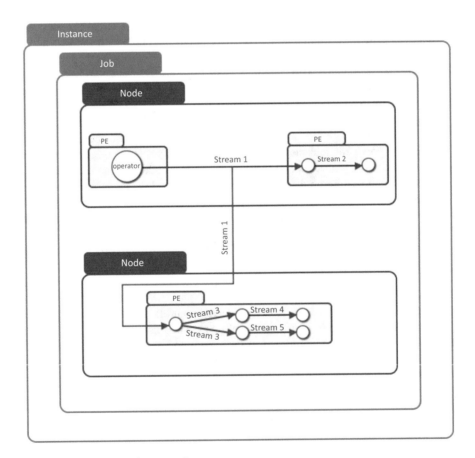

Fig. 5.5 InfoSphere runtime execution

transforms some input streams into output streams. In SPL programs, operator invocation implements the specific use of an operator, with specific assigned input and output streams and with locally specified parameters and logic. Each operator invocation names the input and output streams.

5.5 Other Big Stream Processing Systems

The *Stormy* system [156] has been presented as a distributed stream processing service for continuous data processing that relies on techniques from existing Cloud storage systems that are adapted to efficiently execute streaming workloads. It uses distributed hash tables (DHT) [157] to distribute queries across all nodes, and route events from query to query according to the query graph. To achieve high availability, Stormy uses a replication mechanism where queries are replicated on several nodes, and events are concurrently executed on all replicas. As Stormy's architecture is by design decentralized, there is no single point-of-failure. However, Stormy uses the assumption that a query can be completely executed on one node. Thus, there is an upper limit on the number of incoming events of a stream.

Apache S4 (Simple Scalable Streaming System) is another project that has been presented as a general-purpose, distributed, scalable, partially fault-tolerant, and pluggable platform that provides the programmers with the ability to develop stream data processing applications [158]. The design of S4 is inspired by both of the MapReduce framework [20] and the IBM System S [159]. The S4 architecture provides location and encapsulation transparency which allows applications to be concurrent while providing a simple programming model to application developers. In S4, processing elements (PEs) are the basic computational units and keyed data events are routed with affinity to processing elements (PEs), which consume the events and do one or both of the following: emit one or more events which may be consumed by other PEs, or publish results [158]. In addition, all nodes in the cluster are identical and there is no centralized control. The PEs in the S4 system are assigned to processing nodes which accept input events and then dispatching them to PEs running on them. The events in the S4 are routed to processing nodes by doing a hashing function on the key attribute values of events. So events with the same value for the key attributes are always routed to the same processing node. The coordination between the processing nodes and the messaging between nodes happens through a communication layer which is managed by Apache ZooKeeper. However, one of the main limitations of the runtime of Apache S4 is the lack of reliable data delivery which can be a problem for several types of applications. In S4, applications are written in terms of small execution units (or processing elements) which are designed to be reusable, generic, and configurable so that it can be utilized by various applications.

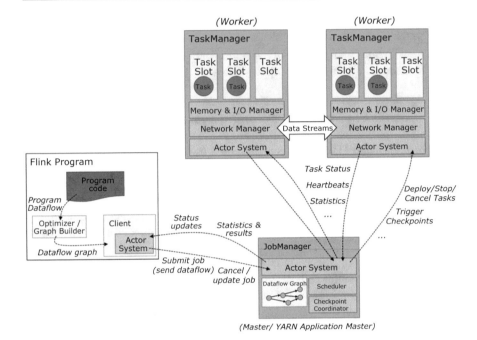

Fig. 5.6 Flonk's architecture

Flink Streaming[10] is an extension of the core Flink API for high-throughput and low-latency data stream processing. The system can connect to and process data streams from various data sources (e.g., Flume, ZeroMQ) where data streams can be transformed and modified using high-level functions that are similar to the ones provided by the batch processing API. The architecture of Flink follows the common master/slave architecture with three main components (Fig. 5.6):

- *Job Manager*: It represents the coordinator node (master node) of the distributed execution that manages the dataflow over the task managers on the slave nodes.
- *Task Manager*: It is responsible for executing the operators that receive and produce streams, delivering their status to Job Manager, and exchanging the data streams between the operators (task managers).
- *Client*: It transforms the program code to a dataflow graph which is submitted to the `Job Manager` for execution.

The system can connect to and process data streams from various data sources (e.g., Kafka, Flume, ZeroMQ) where data streams can be transformed using a set of data

---

[10]http://ci.apache.org/projects/flink/flink-docs-release-0.7/streaming_guide.html.

Fig. 5.7 Spark streaming

stream-based APIs.[11] Flink supports APIs for different programming languages including Java, Scala, and Python. Flink can run as a completely independent framework but can run on top of HDFS and YARN as well. Flink can also directly allocate hardware resources from infrastructure-as-a-service clouds, such as Amazon EC2.

The Spark project has been introduced as a general-purpose Big Data processing engine that can be used for many types of data processing scenarios [44]. The fundamental programming abstraction of Spark is called resilient distributed datasets (RDD) which represents a logical collection of data partitioned across machines that are created by referencing datasets in external storage systems, or by applying various and rich coarse-grained transformations (e.g., Map, Filter, Reduce, Join) on existing RDDs. The Spark system provided an API extension that adds support for continuous stream processing. Initially, Spark streaming relied on the *micro-batch* processing mechanism which collects all data that arrive within a certain period of time and runs a regular batch program on the collected data. In particular, Spark provided the *DStream* API[12] as a Spark streaming abstraction over RDDs where RDDs in a DStream contain data of a given batch interval. During the execution of the batch task, the dataset for the next mini-batch is gathered. Therefore, it can be considered as a batch processing mechanism with controlled window time for stream processing (Fig. 5.7). Recently, from version 2.2.0, Spark provided the new structured streaming feature[13] that supports real-time processing instead of micro-batches [160]. In particular, Spark Structured Steaming manages streaming data as of a relational table where data is continuously appended to this table. Thus, for the end user, the programming interfaces of Structured Steaming is similar to the ones of batch processing while Spark hides and transparently manages all the details of continuous processing of streaming data.

Apache Kafka[14] has been introduced by Linkedin, in 2011, as a fault-tolerant distributed messaging system [161]. In particular, it provides low-latency, high-

---

[11]https://ci.apache.org/projects/flink/flink-docs-release-1.2/dev/datastream_api.html.

[12]https://spark.apache.org/docs/2.2.1/api/java/org/apache/spark/streaming/dstream/DStream.html.

[13]https://spark.apache.org/docs/latest/structured-streaming-programming-guide.html.

[14]https://kafka.apache.org/.

throughput publish and subscribe pipelines. Kafka is mostly used for applications that transform or react to the data streams with real-time requirements. It provides reliable data pipelines for data transfers among applications. In 2016, `Kafka Streams`[15] has been introduced as a new component of Kafka which is designed to support building applications that transform Kafka input stream of topics to Kafka output stream of topics in a distributed and fault-tolerant way. It supports windowing, join and aggregation operations on event-time processing. It is designed as a library and thus it does not have any external dependency and does not require a dedicated cluster environment. In practice, streaming applications use the Kafka Streams library with its embedded a high-level DSL or its API depending on the application's needs.

Apache Samza[16] is another distributed stream processing platform that was originally developed by LinkedIn and then donated to Apache Software Foundation. Samza uses Apache Yarn[17] for distributed resource allocation and scheduling. It also uses Apache Kafka as its distributed message broker. Samza provides an API for creating and running stream tasks on a cluster managed by Yarn. The system is optimized for handling large messages and provides file system persistence for messages. In Kafka a stream is called a Topic and topics are distributed across the brokers using a partitioning scheme. The partitioning of a stream is done based on a key associated with the messages in the stream. The messages with the same key will always belong to the same partition. Samza uses the topic partitioning of Kafka to achieve the distributed partitioning of streams.

`Hazelcast Jet`[18] has been introduced as a new stream processing engine which is built on top of the Hazelcast IMDG (In-Memory Data Grid).[19] It has been designed as a lightweight library that can be embedded in any application to manage a data processing microservice. The library provides APIs that support several operations including filter, group, and map. To model the application, Jet uses directed acyclic graphs (DAGs) where nodes represent computation steps. These computation steps can be executed in parallel by more than one instances of the streaming processor. Vertices are connected with each other via edges that represent the flow of the data and describe how it is routed from the source vertex to the downstream node. They are implemented in a way to buffer the data produced by an upstream node and then let the downstream vertex to pull it. Thus, there are always concurrent queues running among processor instances and they are completely wait-free. The main focus of Hazelcast Jet is to achieve high performance. Thus, it relies on the use of cooperative multi-threading, and, thus, instead of the operating system, the Jet engine is the one which decides how many tasks or threads to run depending on available cores during the runtime. Regarding connectors, for now, Hazelcast Jet

---

[15] https://kafka.apache.org/documentation/streams/.

[16] http://samza.apache.org/.

[17] http://hadoop.apache.org/docs/current/hadoop-yarn/hadoop-yarn-site/YARN.html.

[18] https://hazelcast.com/products/jet/.

[19] http://docs.hazelcast.org/docs/latest-dev/manual/html-single/index.html.

only supports Hazelcast IMDG, where HDFS and Kafka libraries are being actively developed.

Apache Heron[20] has been introduced by Twitter in 2015 as a successor and re-implementation of the `Storm` project with better performance, lower resource consumption, better scalability, and improvement of different architectural components including the job scheduler [162]. In particular, Heron is designed to run on top of YARN, `Mesos`, and ECS (`Amazon EC2 Docker Container Service`). This is a main difference with Storm as Storm relies on the master node, *Nimbus*, as an integral component that is responsible for scheduling management. Heron's APIs have been designed to be fully compatible with Storm APIs to ease the application migration process from Storm to Heron. *JStorm*[21] is another distributed and fault-tolerant real-time stream computation system that has been presented by Alibaba. JStorm represents a complete rewrite of the original `Storm` engine in Java with the promise of better performance.

AthenaX[22] has been released by `Uber Technologies` as a streaming analytics platform that supports building streaming application using Structured Query Language (SQL). `Apache Apex`[23] has been introduced as a Hadoop YARN native platform for stream processing. It provides a library of operators that help the end user to build their streaming applications. It also provides many connectors for messaging systems, databases, and file systems.

Apache Beam (Batch + strEAM)[24] has been created by Google to provide a unified programming model for defining and executing data processing pipelines. In particular, once the application logic is written, the user can choose one of the available runners (e.g., Apache Flink, Apache Spark, Apache Samza) to execute the application workflow. In Beam, a pipeline encapsulates the workflow of your entire data processing tasks from start to end including reading input data, transforming that data, and writing output data.

Apache SAMOA (Scalable Advanced Massive Online Analysis)[25] has been introduced as a platform for mining big data streams [163]. It provides a collection of distributed streaming algorithms for the most common data mining and machine learning tasks such as classification, clustering, and regression, as well as programming abstractions to develop new algorithms. The framework is designed to allow running on several distributed stream processing engines such as Storm, Flink, S4, and Samza.

---

[20]https://apache.github.io/incubator-heron/.

[21]http://jstorm.io/.

[22]https://athenax.readthedocs.io/.

[23]https://apex.apache.org/.

[24]https://beam.apache.org/.

[25]https://samoa.incubator.apache.org/.

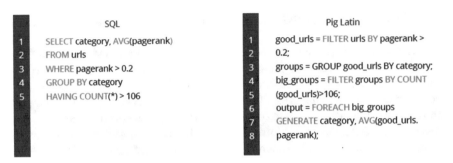

Fig. 5.8 An example SQL query and its equivalent Pig Latin program [164]

5.6   Big Data Pipelining Frameworks

In practice, one of the common use case scenarios is that users need to be able to break down the barriers between data silos such that they are able to design computations or analytics that combine different types of data (e.g., structured, unstructured, stream, graph) or jobs. In order to tackle this challenge, several frameworks have been introduced to build pipelines of big data processing jobs.

5.6.1   Pig Latin

Olston et al. [164] have presented a language called *Pig Latin* that takes a *middle* position between expressing task using the high-level declarative querying model in the spirit of SQL and the low-level/procedural programming model using MapReduce. Pig Latin is implemented within the scope of the *Apache Pig* project[26] and has been used by programmers at Yahoo! for developing data analysis tasks. Writing a Pig Latin program is similar to specifying a query execution plan (e.g., a dataflow graph). To experienced programmers, this method is more appealing than encoding their task as an SQL query and then coercing the system to choose the desired plan through optimizer hints. In general, automatic query optimization has its limits especially when used with uncataloged data, prevalent user-defined functions, and parallel execution, which are all features of the data analysis tasks targeted by the MapReduce framework. Figure 5.8 shows an example SQL query and its equivalent Pig Latin program. Given a URL table with the structure $(url, category, pagerank)$, the task of the SQL query is to find each large category and its average pagerank of high-pagerank urls (>0.2). A Pig Latin program is described as a sequence of steps where each step represents a single data transformation. This characteristic is appealing to many programmers. At the

---

[26]http://incubator.apache.org/pig.

same time, the transformation steps are described using high-level primitives (e.g., filtering, grouping, aggregation) much like in SQL.

Pig Latin has several other features that are important for casual ad-hoc data analysis tasks. These features include support for a flexible, fully nested data model, extensive support for user-defined functions, and the ability to operate over plain input files without any schema information [165]. In particular, Pig Latin has a simple data model consisting of the following four types:

1. *Atom*: An atom contains a simple atomic value such as a string or a number, e.g., "alice."
2. *Tuple*: A tuple is a sequence of fields, each of which can be any of the data types, e.g., ("alice," "lakers").
3. *Bag*: A bag is a collection of tuples with possible duplicates. The schema of the constituent tuples is flexible where not all tuples in a bag need to have the same number and type of fields

    e.g., $\left\{ \begin{array}{l} \text{("alice," "lakers")} \\ \text{("alice," ("iPod," "apple"))} \end{array} \right\}$

4. *Map*: A map is a collection of data items, where each item has an associated key through which it can be looked up. As with bags, the schema of the constituent data items is flexible However, the keys are required to be data atoms, e.g.

    $\left\{ \begin{array}{l} \text{"k1"} \rightarrow \text{("alice", "lakers")} \\ \text{"k2"} \rightarrow \text{"20"} \end{array} \right\}$

To accommodate specialized data processing tasks, Pig Latin has extensive support for user-defined functions (UDFs). The input and output of UDFs in Pig Latin follow its fully nested data model. Pig Latin is architected such that the parsing of the Pig Latin program and the logical plan construction are independent of the execution platform. Only the compilation of the logical plan into a physical plan depends on the specific execution platform chosen. Currently, Pig Latin programs are compiled into sequences of MapReduce jobs which are executed using the Hadoop MapReduce environment. In particular, a Pig Latin program goes through a series of transformation steps [164] before being executed as depicted in Fig. 5.9. The parsing steps verifies that the program is syntactically correct and that all referenced variables are defined. The output of the parser is a canonical logical plan with a one-to-one correspondence between Pig Latin statements and logical operators which are arranged in a directed acyclic graph (DAG). The logical plan generated by the parser is passed through a logical optimizer. In this stage, logical optimizations such as projection pushdown are carried out. The optimized logical plan is then compiled into a series of MapReduce jobs which are then passed through another optimization phase. The DAG of optimized MapReduce jobs is then topologically sorted and jobs are submitted to Hadoop for execution.

Fig. 5.9 The compilation and execution steps of Pig programs [164]

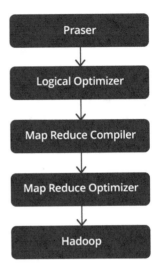

5.6.2 Tez

Apache Tez[27] is another generalized data processing framework [65]. It allows building dataflow driven processing runtimes by specifying complex directed acyclic graph of tasks for high performance batch and interactive data processing applications (Fig. 5.10). In particular, Tez is a client-side application that leverages YARN local resources and distributed cache so that there is no need for deploying any additional components on the underlying cluster. In Tez,[28] data processing is represented as a graph with the vertices in the graph representing processing of data and edges representing movement of data between the processing. Tez uses an event-based model to communicate between tasks and the system, and between various components. These events are used to pass information such as task failures to the required components whereby the flow of data is from output to the input such as the location of data that it generates.

In Tez, the output of a MapReduce job can be directly streamed to the next MapReduce job without writing to HDFS. If there is any failure, the tasks from the last checkpoint will be executed. The data movement between the vertices can happen in-memory, streamed over the network or written to the disk for the sake of checkpointing. In principle, Tez is data type agnostic so that it is only concerned with the movement of data and not with the structure of data format (e.g., key-value pairs, tuples, csv). In particular, the Tez project consists of three main components: an API library that provides the DAG and Runtime APIs and other client-side libraries to build applications, an orchestration framework that has

---

[27]https://tez.apache.org/.

[28]Tez means Speed in Hindi.

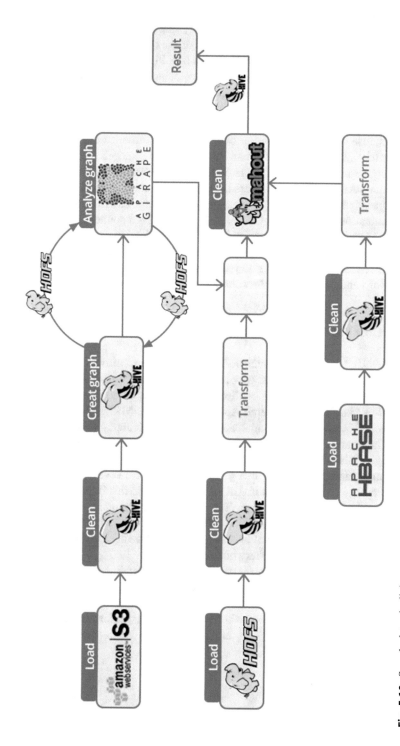

Fig. 5.10 Sample data pipelining

been implemented as a YARN Application Master [66] to execute the DAG in a Hadoop cluster via YARN, and a runtime library that provides implementations of various inputs and outputs that can be used out of the box [65].

In general, Tez is designed for frameworks like Pig, Hive, and not for the end users to directly write application code to be executed by Tez. In particular, using Tez along with Pig and Hive, a single Pig Latin or HiveQL script will be converted into a single Tez job and not as a DAG of MapReduce jobs. However, the execution of a DAG of MapReduce jobs on a Tez can be more efficient than its execution by Hadoop because of Tez's application of dynamic performance optimizations that uses real information about the data and the resources required to process it. The Tez scheduler considers several factors on task assignments including task-locality requirements, total available resources on the cluster, compatibility of containers, automatic parallelization, priority of pending task requests, and freeing up resources that the application cannot use anymore. It also maintains a connection pool of pre-warmed JVMs with shared registry objects. The application can choose to store different kinds of pre-computed information in those shared registry objects so that they can be reused without having to recompute them later on, and this shared set of connections and container-pool resources can run those tasks very fast.

5.6.3 Other Pipelining Systems

Apache MRQL[29] is another framework that has been introduced as a query processing and optimization framework for distributed and large-scale data analysis. MRQL started at the University of Texas at Arlington as an academic research project and in March 2013 it entered Apache Incubation. MRQL has been built on top of Apache Hadoop, Spark, Hama, and Flink. In particular, it provides an SQL-like query language that can be evaluated in four *independent* modes: MapReduce mode using Apache Hadoop, Spark mode using Apache Spark, BSP mode using Apache Hama, and Flink mode using Apache Flink. However, we believe that further research and development is still required to tackle this important challenge and facilitate the job of the end users. The MRQL query language provides a rich type system that supports hierarchical data and nested collections uniformly. It allows nested queries at any level and at any place allowing it to operate on the grouped data using OQL and XQuery queries.

Apache Crunch[30] is a Java library for implementing pipelines that are composed of many user-defined functions and can be executed on top of Hadoop and Spark engines. Apache Crunch is based on Google's FlumeJava library [166] and is efficient for implementing common tasks like joining data, performing aggregations, and sorting records. *Cascading*[31] is another software abstraction layer for the

---

[29]https://mrql.incubator.apache.org/.

[30]https://crunch.apache.org/.

[31]http://www.cascading.org/.

Table 5.1 Feature summary of pipelining frameworks

	Pig	Tez	MRQL	Crunch	Cascading	Pipeline61
Data model	Schmea	Vertex	Schema	PCollection	Pipe	Flexible
Language	Pig Latin	Java	SQL	Java	Scala/Java	Scala/Java
Pipes	Queries	Edges	SQL	Operators	Operators	Flexible Pipes
Engines	Hadoop	Hadoop	Hadoop, Spark Flink, Hama	Hadoop, Spark	Hadoop	Hadoop, Spark Scripts

Hadoop framework which is used to create and execute data processing workflows on a Hadoop cluster using any JVM-based language (Java, JRuby, Clojure, etc.), hiding the underlying complexity of the Hadoop framework.

Pipeline61 [167] has been presented as a framework that supports the building of data pipelines involving heterogeneous execution environments. Pipeline61 is designed to reuse the existing code of the deployed jobs in different environments and also provides version control and dependency management that deals with typical software engineering issues. In particular, Pipeline61 integrates data processing components that are executed on various environments, including MapReduce, Spark, and Scripts. The architecture of Pipeline61 consists of three main components within the framework: Execution Engine, Dependency and Version Manager, and Data Service. The Execution Engine is responsible for triggering, monitoring, and managing the execution of pipelines. Data Service provides a uniformly managed data I/O layer that manages the tedious work of data exchange and conversion between various data sources and execution environments. The Dependency and Version Manager provides a few mechanisms to automate the version control and dependency management for both data and components within the pipelines. Table 5.1 summarizes and compares the features of the various pipelining frameworks.

Large-Scale Machine/Deep Learning Frameworks

6

6.1   Harnessing the Value of Big Data

The continuous growth and integration of data storage, computation, digital devices, and networking empowered a rich environment for the explosive growth of Big Data as well as the tools through which data is produced, shared, cured, and analyzed. In addition to the 3Vs characteristics (Volume, Velocity, Variety and Veracity), it is vital to consider an additional feature of Big Data that is *Value*. In practice, harnessing the value of big data is achieved by analyzing Big Data and extracting from them hidden patterns, trends, and knowledge models by using smart data analysis algorithms and techniques. In general, machine learning and deep learning techniques have been acknowledged as effective means to analyze Big Data and harness the value of data by making it possible to better understand phenomena and behaviors, optimizing processes, and improving machine, business, and scientific discovery. In particular, machine learning and deep learning techniques enable discovering correlations and patterns via analyzing massive amounts of data from various sources that are of different types. In principle, machine learning and deep learning techniques cannot be considered as new techniques. They are originated back to the 1960s. However, recently, they have been gaining increasing momentum and popularity due to the wide availability of massive amounts of datasets in addition to the increasing availability of advanced computing capacities and processing power. Therefore, recently, several frameworks have been developed to support executing machine learning and deep learning algorithms using big data processing frameworks [168].

© The Editor(s) (if applicable) and The Author(s), under exclusive
licence to Springer Nature Switzerland AG 2020
S. Sakr, *Big Data 2.0 Processing Systems*,
https://doi.org/10.1007/978-3-030-44187-6_6

6.2   Big Machine Learning Frameworks

Machine learning [169] is described as the field of study that gives computers the ability to learn without being explicitly programmed. Nowadays, machine learning techniques and algorithms are employed in almost every application domain (e.g., financial applications, advertising, recommendation systems, user behavior analytics). They are playing a crucial role in harnessing the power of the Big Data which we are currently producing every day in our digital Era. In general, machine learning techniques can be generally classified into: *supervised learning* techniques (e.g., classification, regression) and *unsupervised learning* techniques (e.g., clustering). In supervised learning, a training set of examples with the correct labels is provided and, based on this training set, the algorithm generalizes to respond correctly to all possible inputs. This is also called learning from exemplars. In unsupervised learning, correct labels are not provided, but instead the algorithm tries to identify similarities between the inputs so that inputs that have something in common are categorized together. The statistical approach to unsupervised learning is known as density estimation.

Apache Mahout [170] is an open-source toolkit which has pioneered tackling the challenging of implementing scalable machine learning algorithms on top of big data processing frameworks (e.g., Hadoop). Thus, *Mahout*[1] is primarily meant for distributed and batch processing of massive sizes of data on a cluster. In particular, Mahout is essentially a set of Java libraries which is well integrated with Apache Hadoop and is designed to make machine learning applications easier to build. Recently, Mahout has been extended to provide support for machine learning algorithms for collaborative filtering and classification on top of Spark and H2O platforms. *MLlib* [171] has been presented as the Spark's [44] distributed machine learning library that is well-suited for iterative machine learning tasks. MLlib[2] provides scalable implementations of standard learning algorithms for common learning settings including classification, regression, collaborative filtering, clustering, and dimensionality reduction. MLlib supports several languages (e.g., *Java*, *Scala*, and *Python*) and provides a high-level API that leverages Spark's rich ecosystem to simplify the development of end-to-end machine learning pipelines.

H2O[3] is an open-source framework that provides a parallel processing engine which is equipped with math and machine learning libraries. It offers support for various programming languages including Java, R, Python, and Scala. The machine learning algorithms are implemented on top of the H2O's distributed MapReduce framework and exploit the Java Fork/Join framework for implementing multi-threading. H2O implements many machine learning algorithms, such as generalized linear modeling (e.g., linear regression, logistic regression), Naïve Bayes, principal components analysis (PCA), time series, K-means clustering, neural networks, and

---

[1] https://mahout.apache.org/.

[2] https://spark.apache.org/mllib/.

[3] http://www.h2o.ai.

others. H2O also implements complex data mining strategies such as Random Forest, Gradient Boosting, and Deep Learning. Users on H2O can build thousands of models and compare them to get the best prediction results. H2O runs on a several cloud platforms, including `Amazon EC2` and `S3 Storage`, `Microsoft Azure` and `IBM DSX`.[4]

SciDB [172] has been introduced as an analytical database which is oriented toward the data management needs of scientific workflows. In particular, it mixes statistical and linear algebra operations with data management operations using a multi-dimensional array data model. SciDB supports both a functional (AFL) and an SQL-like query language (AQL) where AQL is compiled into AFL. *MLog* [173] has been presented as a high-level language that integrates machine learning into data management systems. It extends the query language over the SciDB data model [174] to allow users to specify machine learning models in a way similar to traditional relational views and relational queries. It is designed to manage all data movement, data persistence, and machine-learning related optimizations automatically. The data model of MLog is based on tensors instead of relations. In fact, all data in MLog are tensors and all operations are a subset of linear algebra over tensors.

MADlib [175] provided a suite of SQL-based implementation for data mining and machine learning algorithms that are designed to get installed and run at scale within any relational database engine that supports extensible SQL, with no need for data import/export to other external tools. The analytics methods in MADlib are designed both for in- or out-of-core execution, and for the shared-nothing scale-out parallelism provided by modern parallel database engines, ensuring that computation is done near to the data. The core functionality of MADlib is written in declarative SQL statements, which orchestrate data movement to and from disk, and across networked computers.

R^5 is currently considered as the de-facto standard in statistical and data analytics research. It is the most popular open-source and cross platform software which has very wide community support. It is flexible, extensible, and comprehensive for productivity. *R* provides a programming language which is used by statisticians and data scientists to conduct data analytics tasks and discover new insights from data by exploiting techniques such as clustering, regression, classification, and text analysis. It is equipped with very rich and powerful library of packages. In particular, *R* provides a rich set of built-in as well as extended functions for data extraction, data cleaning, data loading, data transformation, statistical analysis, machine learning, and visualization. In addition, it provides the ability to connect with other languages and systems (e.g., *Python*). In practice, a main drawback with *R* is that most of its packages were developed primarily for in-memory and interactive usage, i.e., for scenarios in which the data fit in memory. With the aim of tackling this challenge and providing the ability to handle massive datasets, several systems have

---

[4]https://datascience.ibm.com/.

[5]https://www.r-project.org/.

been developed to support the execution of R programs on top of the distributed and scalable Big Data processing platforms such as *Hadoop* (e.g., *Ricardo* [176], *RHadoop*[6] and *RHIPE*,[7] *Segue*[8]) and *Spark* [44] (e.g., *SparkR* [177]). For example, *RHIPE* is an R package that brings MapReduce framework to R users and enables them to access the Hadoop cluster from within the R environment. In particular, by using specific R functions, users are able to launch MapReduce jobs on the Hadoop cluster where the results can be easily retrieved from HDFS. *Segue* enables users to execute MapReduce jobs from within the R environment on Amazon Elastic MapReduce platforms. *SparkR* has become a popular *R* package that supports a lightweight front-end to execute *R* programs on top of the *Apache Spark* [44] distributed computation engine and allows executing large-scale data analysis tasks from the *R* shell. *Pydoop* [178] is a Python package that provides an API for both the Hadoop framework and the HDFS.

Google has provided a cloud-based SaaS machine learning platform[9] which is equipped with pre-trained models in addition to a platform to generate users' models. The service is integrated with other Google services such as Google Cloud Storage and Google Cloud Dataflow. It encapsulates powerful machine learning models that support different analytics applications (e.g., image analysis, speech recognition, text analysis, and automatic translation) through REST API calls. Similarly, Amazon provides its machine learning as a service solution[10] (AML) which guides its users through the process of creating data analytics models without the need to learn complex algorithms or technologies. Once the models are created, the service makes it easy to perform predictions via simple APIs without the need to write any user code or manage any hardware or software infrastructure. AML works with data stored in *Amazon S3, RDS*, or *Redshift*. It provides an API set for connecting with and manipulating other data sources. AML relies on Amazon SageMaker platform that allows the user to build, train, and deploy their machine learning models. *IBM Watson Analytics*[11] is another SaaS predictive analytic framework that allows its user to express their analytics job using natural English language. The service attempts to automatically spot interesting correlations and exceptions within the input data. It also provides suggestions on the various data cleaning steps and the adequate data visualization technique to use for various analysis scenarios.

Microsoft introduced *AzureML* [179] as a machine learning framework solution which provides a cloud-based visual environment for constructing data analytics workflows. Azure ML is often described as a SaaS; however, it can be seen also a PaaS since it can be used develop SaaS solutions on top of it. It is provided as a fully

---

[6]https://github.com/RevolutionAnalytics/RHadoop.

[7]https://github.com/tesseradata/RHIPE.

[8]https://code.google.com/archive/p/segue/.

[9]https://cloud.google.com/products/machine-learning/.

[10]https://aws.amazon.com/machine-learning/.

[11]https://www.ibm.com/analytics/watson-analytics/.

managed service by Microsoft where users neither need to buy any hardware/software nor manually manage any virtual machines. AzureML provides data scientists with a web-based machine learning IDE for creating and automating machine learning workflows. In addition, it provides scalable and parallel implementations of popular machine learning techniques as well as data processing capabilities using a drag-and-drop interface. AzureML can read and import data from various sources including HTTP URL, Azure Blob Storage, Azure Table, and Azure SQL Database. It also allows data scientists to import their own custom data analysis scripts (e.g., in R or Python). *Cumulon* [180] has been present as a system which is designed to help users rapidly develop and deploy matrix-based big-data analysis programs in the cloud. It provides an abstraction for distributed storage of matrices on top of HDFS. In particular, matrices are stored and accessed by tiles. A Cumulon program executes as a workflow of jobs. Each job reads a number of input matrices and writes a number of output matrices; input and output matrices must be disjoint. Dependencies among jobs are implied by dependent accesses to the same matrices. Dependent jobs execute in serial order. Each job executes multiple independent tasks that do not communicate with each other. Hadoop-based Cumulon inherits important features of Hadoop such as failure handling and is able to leverage the vibrant Hadoop ecosystem. While targeting matrix operations, Cumulon can support programs that also contain traditional, non-matrix Hadoop jobs.

The *BigML*[12] SaaS framework supports discovering predictive models from the input data using data classification and regression algorithms. In BigML, predictive models are presented to the users as an interactive decision tree which is dynamically visualized and explored within the BigML interface. BigML also provides a PaaS solution, *BigML PredictServer*,[13] which can be integrated with applications, services, and other data analysis tools. *Hunk*[14] is a commercial data analysis platform developed for rapidly exploring, analyzing, and visualizing data in Hadoop and NoSQL data stores. Hunk uses a set of high-level user and programming interfaces to improve the speed and simplicity of getting insights from large unstructured and structured datasets. One of the key components of the Hunk architecture is the Splunk Virtual Index. This system decouples the storage tier from the data access and analytics tiers, so enabling Hunk to route requests to different data stores. The analytics tier is based on Splunk's Search Processing Language (SPL) that is designed for data exploration across large, different datasets. The Hunk web framework allows building applications on top of the Hadoop Distributed File System (HDFS) and/or the NoSQL data store. Developers can use Hunk to build their Big Data applications on top of the data in Hadoop using a set of well-known languages and frameworks. Indeed, the framework enables developers to integrate data and functionality from Hunk into enterprise Big Data applications using a web

---

[12]https://bigml.com.

[13]https://bigml.com/predictserver.

[14]http://www.splunk.com/en_us/products/hunk.html.

framework, documented REST API, and software development kits for C#, Java, JavaScript, PHP, and Ruby.

Several declarative machine learning implementations have been implemented on top of Big Data processing systems [181]. For example, *Samsara* [182] has been introduced as a mathematical environment that supports declarative implementation for general linear algebra and statistical operations as part of the Apache Mahout library. It allows its users to specify programs in an R-like style using a set of common matrix abstractions and linear algebraic operations. Samsara compiles, optimizes, and executes its programs on distributed dataflow systems (e.g., *Apache Spark, Apache Flink, H2O*). *MLbase* [183] has been implemented to provide a general-purpose machine learning library with a similar goal to Mahout's goal to provide a viable solution for dealing with large-scale machine learning tasks on top of the Spark framework. It supports a Pig Latin-like [164] declarative language to specify machine learning tasks and implements and provides a set of high-level operators that enable its users to implement a wide range of machine learning methods without deep systems knowledge. In addition, it implements an optimizer to select and dynamically adapt the choice of learning algorithm.

Apache SystemML[15] provides a declarative machine learning framework which is developed to run on top of Apache Spark [181]. It supports R and Python-like syntax that includes statistical functions, linear algebra primitives, and ML-specific constructs. It applies cost-based compilation techniques to generate efficient, low-level execution plans with in-memory single-node and large-scale distributed operations.

ScalOps [184] has been presented as a domain-specific language (DSL) that moves beyond single pass data analytics (i.e., MapReduce) to include multi-pass workloads, supporting iteration over algorithms expressed as relational queries on the training and model data. The physical execution plans of ScalOps consist of dataflow operators which are executed using the Hyracks data-intensive computing engine [185].

Keystoneml framework [186] has been designed to support building complex and multi-stage pipelines that include feature extraction, dimensionality reduction, data transformations, and training supervised learning models. It provides a high-level, type-safe API that is built around logical operators to capture end-to-end machine learning applications. To optimize the machine learning pipelines, Keystoneml applies techniques to do both per-operator optimization and end-to-end pipeline optimization. It uses a cost-based optimizer that accounts for both computation and communication costs. The optimizer is also able to determine which intermediate states should be materialized in the main memory during the iterative execution over the raw data.

---

[15]https://systemml.apache.org/.

6.3   Deep Learning Frameworks

Most machine learning algorithms only have the ability to use one or two layers of data transformation to learn the output representation. Thus, they are called as shallow models. As datasets continue to grow in the dimensions of the feature space, finding the optimal output representation with a shallow model is not always possible. Deep learning tackles this challenge by providing a multi-layer approach to learn data representations, typically performed with a multi-layer neural network. In general, deep learning [187, 188] is considered as a sub-field of machine learning techniques which is gaining increasing popularity due to its success in various application domains including natural language processing, medical diagnosis, speech recognition, and computer vision [189–191]. In general, deep learning techniques represent a subset of machine learning methodologies that are based on artificial neural networks (ANN) which are mainly inspired by the neuron structure of the human brain [192]. It is described as *deep* because it has more than one layer of nonlinear feature transformation (Fig. 6.1). In practice, the main advantage of deep learning over the traditional machine learning techniques is their ability for automatic feature extraction which allows learning complex functions to be mapped from the input space to the output space without much human intervention. In particular, it consists of multiple layers, nodes, weights, and optimization algorithms. Due to the increasing availability of labeled data,

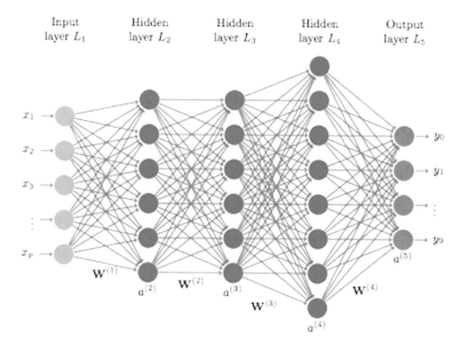

Fig. 6.1 Deep learning architecture

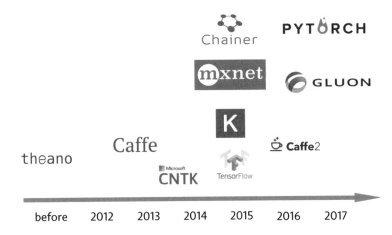

Fig. 6.2 Timeline of deep learning frameworks

computing power, better optimization algorithms, and better neural net models and architectures, deep learning techniques have started to outperform humans in some domains such as image recognition and classification (Fig. 6.2).

TensorFlow[16] is an open-source library for high performance computation and large-scale machine learning across different platforms including CPU, GPU, and distributed processing [193]. TensorFlow is developed by Google Brain team in Google's AI organization and has been released as an open-source project in 2015. It provides a dataflow model that allows mutable state and cyclic computation graph. TensorFlow supports different types of architecture due to its auto differentiation and parameter sharing capabilities. TensorFlow supports parallelism through parallel execution of dataflow graph model using multiple computational resources that collaborate to update shared parameters. The computation in TensorFlow is modeled as a directed graph where nodes represent operations. Values that flow along the edges of the graph are called *Tensors* that are represented as a multi-dimensional array. An operation can take zero or more tensors as input and produce zero or more tensors as output. An operation is valid as long as the graph which the operation is part of is valid. The main focus of TensorFlow is to simplify the real-world use of machine learning system and significantly reducing the maintenance burdens. *TFX* [194] has been presented by Google as a TensorFlow-based general-purpose machine learning platform that integrates different components including a learner for generating models based on training data, modules for analyzing and validating both data as well as models in addition to infrastructure for serving models in production.

---

[16]https://www.tensorflow.org/.

Keras[17] is an open-source deep learning framework developed by François Cholle, a member of Google AI team. Keras is considered as a meta-framework that interacts with other frameworks. In particular, it can run on the top of TensorFlow and Theano [195]. It is implemented in Python and provides high-level neural networks APIs for developing deep learning models. Instead of handling low-level operations (differentiation and tensor manipulation), Keras relies on a specialized library that serves as its back-end engine. Keras minimizes the number of actions required by a user for a specific action. An important feature of Keras is its ease of use without scarifying flexibility. Keras enables the users to implement their models as if they were implemented on the base frameworks (such as TensorFlow, Theano, MXNet).

MXNet[18] is an open-source deep learning framework founded as a collaboration between Carnegie Mellon University, Washington University, and Microsoft. It is a scalable framework that allows training deep neural networks using different programming languages including C++, Python, Matlab, JavaScript, R, Julia, and Scala. MXNet supports data parallelism on multiple CPUs or GPUs and allows model parallelism as well. MXNet supports two different modes of training; synchronous and asynchronous training [196]. MXNet provides primitive fault tolerance operations through save and load: save stores the model's parameters to a checkpoint file and load restores the model's parameters from a checkpoint file. MXNet supports both declarative programming and imperative programming.

Theano[19] is an open-source Python library for fast large-scale computations that can run on different computing platforms including CPU and GPU [195]. Theano has been developed by researchers and developers from Montreal university. Theano is a fundamental mathematical expression library that facilitates building deep learning models. Different libraries have been developed on the top of Theano such as Keras which is tailored for building deep learning models and provides the building blocks for efficient experimentation of deep learning models. Computations in Theano are expressed using NumPy-esque syntax. Theano works by creating a symbolic representation of the operations which are translated to C++ and then compiling them into dynamically loaded Python molecules. Theano supports both data parallelism and model parallelism.

PyTorch[20] has been introduced by Facebook's AI research group in October 2016 [197]. PyTorch is a Python-based deep learning framework which facilitates building deep learning models through an easy to use API. Unlike most of the other popular deep learning frameworks, which use static computation graphs, PyTorch uses dynamic computation, which allows greater flexibility in building complex architectures.

---

[17]https://keras.io/.

[18]https://mxnet.apache.org/.

[19]https://deeplearning.net/software/theano//.

[20]https://pytorch.org/.

Chainer[21] is an open-source deep learning framework, implemented in Python. The development of Chainer is led by researchers and developers from Tokyo University [198]. Chainer provides automatic differentiation APIs for building and training neural networks. Chainer's approach is based on the "define-by-run" technique which enables building the computational graph during training and allows the user to change the graph at each iteration. Chainer is a flexible framework as it provides an imperative API in Python and NumPy. Both CPU and GPU computations are supported by Chainer.

---

[21]https://chainer.org/.

Conclusions and Outlook

7

Currently, the world is entirely living the era of the information age [199]. The world is progressively moving toward being a data-driven society where data is the most valuable asset. Therefore, Big data analytics is currently representing a revolution that cannot be missed. It is significantly transforming and changing various aspects in our modern life including the way we live, socialize, think, work, do business, conduct research, and govern society [200]. Over the last few years, data scientist has been recognized as one of the sexiest jobs of the twenty-first century.[1] In practice, efficient and effective analysis and exploitation of big data has become an essential requirement for enhancing the competitiveness of enterprises and maintaining sustained social and economic growth of societies and countries. For example, in the business domain, opportunities for utilizing big data assets to reduce costs, enhance revenue, and manage risks are a representative sample of a long list of useful applications which will continue to expand and grow. In particular, the lag in utilizing big data technologies and applications has become the leading factor that results in a company's loss of business strategic advantages. Therefore, the capacity of any enterprise to accumulate, store, process, and analyze massive amounts of data will represent a new landmark indication for its strength, power, and potential development.

With the emergence of cloud computing technology, the X-as-a-Services (XaaS) model (e.g., IaaS, PaaS, SaaS) has been promoted at different levels and domains [201]. With the emergence of big data processing systems, Analytics as a Service (AaaS) [202] represents a new model which is going to upgrade and transform the big data processing domain. The global giants of the information

---

[1] https://hbr.org/2012/10/data-scientist-the-sexiest-job-of-the-21st-century/.

S. Sakr, *Big Data 2.0 Processing Systems*, https://doi.org/10.1007/978-3-030-44187-6_7

technology business have already begun providing their technical solution that cope with the big data era including as Oracle,[2] Google,[3] Microsoft,[4] and IBM.[5]

In practice, the big data phenomena has invited the scientific research communities to revisit their scientific methods and techniques. Initially, the first paradigm of research methods was mainly based on experiments. The second paradigm of theoretical science was mainly based on the study of various theorems and laws. However, in practice, theoretical analysis turned out to be too complex and not feasible for dealing with various practical problems in many scenarios. Therefore, researchers started to use simulation-based methods which led to the third paradigm of computational science. In 2007, Jim Gray, the Turing Award winner[6] separated data-intensive science from computational science. Gray believed that the fourth paradigm is not only a change in the way of scientific research, but also a change in the way that people think [4]. Recently, Data Science [203] has been gradually emerging as an interdisciplinary discipline which is gaining increasing attention and interest from the research communities. This new discipline spans across many disciplines including computer and information science, statistics, psychology, network science, social science, and system science. It relies on various theories and techniques from several domains including data warehousing, data mining, machine learning, probability theory, statistical learning, pattern recognition, uncertainty modeling, visualization, and high performance computing [204]. As a result, in the recent years, several reputable institutions around the world have been establishing new academic programs,[7] research groups, and research centers which are specialized to serve this new domain and build new generations of talented data engineers and data scientists.

In spite of the high expectations on the promises and potentials of the big data paradigm, there are still many challenges in the way of harnessing its full power [205, 206]. For example, the typical characteristics of big data are diversified types that have complex inter-relationships, and not necessarily consistent high data quality. These characteristics lead to significant increases in computational complexity and the required computing power. Therefore, traditional data processing and analysis tasks including retrieval, mining tasks, sentiment analysis, and semantic analysis become increasingly more complex in comparison to the traditional data. Currently, we lack the fundamental models to understand and quantitatively analyze, estimate, and describe the complexity of big data processing

---

[2]http://www.oracle.com/us/products/database/options/advanced-analytics/overview/index.html.

[3]https://cloud.google.com/bigquery/.

[4]https://www.microsoft.com/en-us/server-cloud/products/analytics-platform-system/overview.aspx.

[5]http://www.ibm.com/analytics/us/en/.

[6]http://amturing.acm.org/award_winners/gray_3649936.cfm.

[7]http://www.mastersindatascience.org/schools/23-great-schools-with-masters-programs-in-data-science/.

jobs. In addition, there is no understanding on the relationship between the data complexity and the computational complexity of big data processing jobs.

In the early days of the Hadoop framework, the lack of declarative languages to express the large-scale data processing tasks has limited its practicality and the wide acceptance and the usage of the framework [22]. Therefore, several systems (e.g., Pig, Hive, Impala, HadoopDB) have been introduced to the Hadoop stack to fill this gap and provide higher-level languages for expressing large-scale data analysis tasks on Hadoop [207]. In practice, these languages have seen wide adoption in the industry and research communities. Currently the systems/stacks of large-scale graph, stream, or pipelining platforms are suffering from the same challenge. Therefore, we believe that it is beyond doubt that, the higher the level of the language abstractions, the easier the user's job for expressing their graph processing jobs [208]. In addition, high-level languages that enable the underlying systems/stack to perform automatic optimization are crucially required and represent an important research direction to enrich this domain.

Processing and analyzing huge data volumes poses various challenges to the design of system architectures and computing frameworks. Even though several systems had been introduced with various design architectures, we are still lacking a deeper understanding of the performance characteristics for the various design architectures in addition to lacking comprehensive benchmarks for the various big data processing platforms. For example, in the domain of benchmarking large-scale graph processing systems, Guo et al. [209] have identified three dimensions of diversity that complicate the process of gaining knowledge and deeper understanding for the performance of graph processing platforms: dataset, algorithm, and platform diversity. Dataset diversity is the result of the wide set of application domains for graph data. Algorithm diversity is an outcome of the different goals of processing graphs (e.g., PageRank, subgraph matching, centrality, betweens). Platform diversity is the result of the wide spectrum of systems which are influenced by the wide diversity of infrastructure (compute and storage systems). To alleviate this challenge and with the crucial need to understand and analyze the performance characteristics of existing big data processing systems, several recent studies have been conducted that attempt to address this challenge [210–216]. For example, Han et al. [212] have conducted another study on *Giraph*, *GPS*, *Mizan*, and *GraphLab* using four different algorithms: PageRank, single source shortest path, weakly connected components, and distributed minimum spanning tree on up to 128 Amazon EC2 machines. The experiments used datasets which are obtained from SNAP[8] (Stanford Network Analysis Project) and LAW[9] (Laboratory for Web Algorithms). The study has considered different metrics for comparison: *total time* which represents the total running time from start to finish and includes both of the *setup time*, the time taken to load and partition the input graph as well as write the output, and *computation time*, which includes local vertex

---

[8]http://snap.stanford.edu/data/.

[9]http://law.di.unimi.it/datasets.php.

computation, barrier synchronization, and communication. In addition, the study has considered the *memory usage* and *total network usage* metrics for its benchmarking. Another benchmarking study has been conducted by Lu et al. [213] to evaluate the performance characteristics of *Giraph, GraphLab/ PowerGraph, GPS, Pregel+*, and *GraphChi*. The study has used large graphs with different characteristics, including skewed (e.g., power-law) degree distribution, small diameter (e.g., small-world), large diameter, (relatively) high average degree, and random graphs. Additionally, the study also used several evaluation algorithms including PageRank, diameter estimation, single source shortest paths (SSSP), and Graph Coloring. Guo et al. [214] have conducted a benchmarking study which considered a set of various systems which are more focused on general-purpose distributed processing platforms. In particular, the study has considered the following systems: *Hadoop YARN*[10] which represents the next generation of Hadoop that separates resource management and job management, *Stratosphere*[11] which is an open-source platform for large-scale data processing [52], *Giraph, GraphLab*, and *Neo4j*[12] which represent one of the popular open-source graph databases. The study has focused on four benchmarking metrics: *raw processing power, resource utilization, scalability*, and *overhead*. In principle, all of these studies have just scratched the surface on the process of evaluating and benchmarking big data processing systems. In practice, we still need to conduct fundamental research with a more comprehensive performance evaluation for the various big data processing systems and architecture including the big stream processing and pipeline framework and the SQL-on-Hadoop systems. We also lack the availability of validation tools, standard benchmarks, and system performance prediction methods that can help to have a deeper and more solid understanding of the strengths and weaknesses of the various big data processing platforms.

In general, more alternatives usually mean harder decisions for choice. In practice, with the wide spectrum of big data processing systems which are currently available, it becomes very challenging to decide by intuition which system is the most adequate for a given application requirements or workload. Making such decision would require significant knowledge about the programming model, design choice, and probably the implementation details of the various available systems [217–220]. In addition, the various benchmarking studies have commonly found out that the performance characteristics of the different systems can widely vary depending on the application workload and there is no single winning system which can always outperform all other systems for the different application workloads. Furthermore, porting the data and the data analytics jobs between different systems is a tedious, time-consuming, and costly task. Therefore, users can become *locked-in* to a specific systems despite the availability of faster or more adequate systems for a given workload. Gog et al. [221] argued that the main reason behind

---

[10]http://hadoop.apache.org/docs/current/hadoop-yarn/hadoop-yarn-site/YARN.html.

[11]http://stratosphere.eu/.

[12]http://neo4j.com/.

this challenge is the tight coupling between user-facing front-ends for implementing the big data jobs and the back-end execution engines that run them. In order to tackle this challenge, they introduced the *Musketeer* system [221] to dynamically map the front-end of the big data jobs (e.g., Hive, SparkSQL) to a broad range of back-end execution engines (e.g., MapReduce, Spark, PowerGraph). *Rheem* [222] is another framework that has been recently introduced to tackle the same challenge by providing platform independence, multi-platform task execution, and features three-layer data processing abstractions. In practice, such emerging frameworks are paving the way for providing the users with more freedom and flexibility in executing their big data analytics jobs in addition to getting way from the low-level optimization details of the various big data processing engines [222].

In practice, Big Data analytics lives and dies by the data. It mainly rests on the availability of massive datasets, of that there can be no doubt. The more data that is available, the richer the insights and the results that Big Data science can produce. The bigger and more diverse the dataset, the better the analysis can model the real world. Therefore, any successful Big Data science process has attempted to incorporate as many datasets from internal and public sources as possible. In reality, data is segmented, siloed, and under the control of different individuals, departments, or organizations. It is crucially required to motivate all parties to work collaboratively and share useful data/insights for the public. Recently, there has been an increasing trend for open data initiatives which supports the idea of making data publicly available to everyone to use and republish as they wish, without restrictions from copyright, patents, or other mechanisms of control [223]. Online data markets [224] are emerging cloud-based services (e.g., `Azure Data Market`,[13] `Kaggle`,[14] `Connect`,[15] `Socrata`[16]). For example, Kaggle is a platform where companies can provide data to a community of data scientists so that they can analyze the data with the aim of discovering predictive, actionable insights and win incentive awards. In particular, such platforms follow a model where data and rewards are traded for innovation. More research, effort, and development are still required in this direction.

In general, a major obstacle for supporting Big Data analytics applications is the challenging and time-consuming process of identifying and training an adequate predictive model. Therefore, data science is a highly iterative exploratory process where most scientists work hard to find the best model or algorithm that meets their data challenge. In practice, there is no one-model-fits-all solutions; thus, there is no single model or algorithm that can handle all dataset varieties and changes in data that may occur over time. All machine learning algorithms require user-defined inputs to achieve a balance between accuracy and generalizability. This task is referred to as parameter tuning. The tuning parameters impact the way the algorithm

---

[13] http://datamarket.azure.com/browse/data.

[14] https://www.kaggle.com/.

[15] https://connect.data.com/.

[16] https://socrata.com/.

searches for the optimal solution. This iterative and explorative nature of the model building process is prohibitively expensive with very large datasets. Thus, recent research efforts have been attempting to automate this process [225]; however, they have mainly focused on single-node implementations and have assumed that model training itself is a black box, limiting their usefulness for applications driven by large-scale datasets [225].

With the increasing number of platforms and services, interoperability is arising as a main issue. Standard formats and models are required to enable interoperability and ease cooperation among the various platforms and services. In addition, the service-oriented paradigm can play an effective role in supporting the execution of large-scale distributed analytics on heterogeneous platforms along with software components developed using various programming languages or tools. Furthermore, in practice, the majority of existing big data processing platforms (e.g., Hadoop and Spark) are designed based on the single-cluster setup with the assumptions of centralized management and homogeneous connectivity which makes them sub-optimal and sometimes infeasible to apply for scenarios that require implementing data analytics jobs on highly distributed datasets (e.g., across racks, clusters, data centers, or multi-organizations). Some scenarios can also require distributing data analysis tasks in a hybrid mode among local processing of local data sources and model exchange and fusion mechanisms to compose the results produced in the distributed nodes.

With the increasing usage of machine learning and the increasing number of models, the issues of model management, model sharing, model versioning, and life-cycle management have become significantly important. For example, it is important to keep track of the models developed and understand the differences between them by recording their metadata (e.g., training sample, hyperparameters). ModelHub [226] has been proposed to provide a model versioning system to store and query the models and their versions, a domain-specific language that serves as an abstraction layer for searching through model space in addition to a hosted service to store developed models, explore existing models, enumerate new models, and share the models with others. ModelDB [227] is another system for managing machine learning models that automatically tracks the models in their native environments (e.g., Mahout, SparkML), indexes them, and allows flexible exploration of models using either SQL or a visual web-based interface. Along with models and pipelines, ModelDB stores several metadata (e.g., parameters of pre-processing steps, hyperparameters for models, etc.) and quality metrics (e.g., AUC, accuracy). In addition, it can store the training and test data for each model.

In practice, building machine learning applications is a highly time-consuming process that requires substantial effort even from best-trained data scientists to deploy, operate, and monitor. One of the main reasons behind this challenge is the lack of tools for supporting end-to-end machine learning application development that can ease and accelerate the job for end users. The DAWN project at Stanford [228] has recently announced its vision for the next 5 years with the aim of making the machine learning (ML) process *usable* for small teams of non-ML experts so that they can easily apply ML to their problems, achieve high-quality results, and deploy production systems that can be used in critical applications. The main design philosophy of the DAWN project is to target the management of end-to-end ML workflows, empower domain experts to easily develop and deploy their models, and perform effective optimization of the workflow execution pipelines using simple interfaces. Another important usability aspect is the *explainability* of the developed models. In general, explainability is very useful for machine learning models used and trained as black boxes where the output models are not easy or intuitive to explain (e.g., SVM, neural networks, deep learning). For example, in the healthcare domain, the physicians should be able to understand and interpret why the developed models are meaningful and applicable. We believe that additional research efforts are crucially required to ease and accelerate the life cycle and the data science process and make it more usable.

References

1. C. Lynch, Big data: how do your data grow? Nature **455**(7209), 28–29 (2008)
2. Large synoptic survey. http://www.lsst.org/
3. H. Chen, R.H.L. Chiang, V.C. Storey, Business intelligence and analytics: from big data to big impact. MIS Q. **36**(4), 1165–1188 (2012)
4. T. Hey, S. Tansley, K. Tolle (eds.), *The Fourth Paradigm: Data-Intensive Scientific Discovery* (Microsoft Research, Redmond, 2009)
5. G. Bell, J. Gray, A.S. Szalay, Petascale computational systems. IEEE Comput. **39**(1), 110–112 (2006)
6. J. Manyika, M. Chui, B. Brown, J. Bughin, R. Dobbs, C. Roxburgh, A.H. Byers, Big data: the next frontier for innovation, competition, and productivity. Technical Report 1999-66, May 2011
7. A. McAfee, E. Brynjolfsson, T.H. Davenport, D.J. Patil, D. Barton, Big data. The management revolution. Harvard Bus. Rev. **90**(10), 61–67 (2012)
8. R. Buyya, C.S. Yeo, S. Venugopal, J. Broberg, I. Brandic, Cloud computing and emerging IT platforms: vision, hype, and reality for delivering computing as the 5th utility. Future Gener. Comput. Syst. **25**(6), 599–616 (2009)
9. L.M. Vaquero, L. Rodero-Merino, J. Caceres, M. Lindner, A break in the clouds: towards a cloud definition. ACM SIGCOMM Comput. Commun. Rev. **39**(1), 50–55 (2008)
10. D.C. Plummer, T.J. Bittman, T. Austin, D.W. Cearley, D.M. Smith, Cloud computing: defining and describing an emerging phenomenon. Gartner (2008)
11. J. Staten, S. Yates, F.E. Gillett, W. Saleh, R.A. Dines, Is cloud computing ready for the enterprise. Forrester Research (2008)
12. M. Armbrust, O. Fox, R. Griffith, A.D. Joseph, Y. Katz, A. Konwinski, G. Lee, D. Patterson, A. Rabkin, I. Stoica et al., Above the clouds: a Berkeley view of cloud computing (2009)
13. S. Madden, From databases to big data. IEEE Internet Comput. **3**, 4–6 (2012)
14. S. Sakr, Cloud-hosted databases: technologies, challenges and opportunities. Clust. Comput. **17**(2), 487–502 (2014)
15. S. Sakr, A. Liu, D.M. Batista, M. Alomari, A survey of large scale data management approaches in cloud environments. IEEE Commun. Surv. Tutor. **13**(3), 311–336 (2011)
16. S. LaValle, E. Lesser, R. Shockley, M.S. Hopkins, N. Kruschwitz, Big data, analytics and the path from insights to value. MIT Sloan Manag. Rev. **52**(2), 21 (2011)
17. X. Wu, X. Zhu, G.-Q. Wu, W. Ding, Data mining with big data. IEEE Trans. Knowl. Data Eng. **26**(1), 97–107 (2014)
18. D.J. DeWitt, J. Gray, Parallel database systems: the future of high performance database systems. Commun. ACM **35**(6), 85–98 (1992)
19. A. Pavlo, E. Paulson, A. Rasin, D.J. Abadi, D.J. DeWitt, S. Madden, M. Stonebraker, A comparison of approaches to large-scale data analysis, in *SIGMOD* (2009), pp. 165–178

S. Sakr, *Big Data 2.0 Processing Systems*, https://doi.org/10.1007/978-3-030-44187-6

20. J. Dean, S. Ghemawa, MapReduce: simplified data processing on large clusters, in *OSDI*, 2004
21. D. Agrawal, S. Das, A. El Abbadi, Big data and cloud computing: current state and future opportunities, in *Proceedings of the 14th International Conference on Extending Database Technology* (ACM, New York, 2011), pp. 530–533
22. S. Sakr, A. Liu, A.G. Fayoumi, The family of MapReduce and large-scale data processing systems. ACM Comput. Surv. **46**(1), 1–44 (2013)
23. H. Yang, A. Dasdan, R. Hsiao, D. Parker, Map-reduce-merge: simplified relational data processing on large clusters, in *SIGMOD*, 2007
24. M. Stonebraker, The case for shared nothing. IEEE Database Eng. Bull. **9**(1), 4–9 (1986)
25. T. White, *Hadoop: The Definitive Guide* (O'Reilly Media, Sebastopol, 2012)
26. D. Jiang, A.K.H. Tung, G. Chen, MAP-JOIN-REDUCE: toward scalable and efficient data analysis on large clusters. IEEE TKDE **23**(9), 1299–1311 (2011)
27. Y. Bu, B. Howe, M. Balazinska, M.D. Ernst, The HaLoop approach to large-scale iterative data analysis. VLDB J. **21**(2), 169–190 (2012)
28. Y. Zhang, Q. Gao, L. Gao, C. Wang, iMapReduce: a distributed computing framework for iterative computation. J. Grid Comput. **10**(1), 47–68 (2012)
29. J. Ekanayake, H. Li, B. Zhang, T. Gunarathne, S.-H. Bae, J. Qiu, G. Fox, Twister: a runtime for iterative MapReduce, in *HPDC*, 2010
30. T. Nykiel, M. Potamias, C. Mishra, G. Kollios, N. Koudas, MRShare: sharing across multiple queries in MapReduce. Proc. VLDB Endowment **3**(1), 494–505 (2010)
31. I. Elghandour, A. Aboulnaga, ReStore: reusing results of MapReduce jobs. Proc. VLDB Endowment **5**(6), 586–597 (2012)
32. I. Elghandour, A. Aboulnaga, ReStore: reusing results of MapReduce jobs in Pig, in *SIGMOD*, 2012
33. J. Dittrich, J.-A. Quiané-Ruiz, A. Jindal, Y. Kargin, V. Setty, J. Schad, Hadoop++: making a yellow elephant run like a cheetah (without it even noticing). Proc. VLDB Endowment **3**(1), 518–529 (2010)
34. A. Floratou, J.M. Patel, E.J. Shekita, S. Tata, Column-oriented storage techniques for MapReduce. Proc. VLDB Endowment **4**(7), 419–429 (2011)
35. Y. Lin et al., Llama: leveraging columnar storage for scalable join processing in the MapReduce framework, in *SIGMOD*, 2011
36. T. Kaldewey, E.J. Shekita, S. Tata, Clydesdale: structured data processing on MapReduce, in *EDBT* (2012), pp. 15–25
37. A. Balmin, T. Kaldewey, S. Tata, Clydesdale: structured data processing on Hadoop, in *SIGMOD Conference* (2012), pp. 705–708
38. M. Zukowski, P.A. Boncz, N. Nes, S. Héman, MonetDB/X100 - a DBMS in the CPU cache. IEEE Data Eng. Bull. **28**(2), 17–22 (2005)
39. Y. He, R. Lee, Y. Huai, Z. Shao, N. Jain, X. Zhang, Z. Xu, RCFile: a fast and space-efficient data placement structure in MapReduce-based warehouse systems, in *ICDE* (2011), pp. 1199–1208
40. A. Jindal, J.-A. Quiane-Ruiz, J. Dittrich, Trojan data layouts: right shoes for a running elephant, in *SoCC*, 2011
41. M.Y. Eltabakh, Y. Tian, F. Özcan, R. Gemulla, A. Krettek, J. McPherson, CoHadoop: flexible data placement and its exploitation in Hadoop. Proc. VLDB Endowment **4**(9), 575–585 (2011)
42. Y. Huai, A. Chauhan, A. Gates, G. Hagleitner, E.N. Hanson, O. O'Malley, J. Pandey, Y. Yuan, R. Lee, X. Zhang, Major technical advancements in Apache Hive, in *SIGMOD*, 2014
43. G. Malewicz, M.H. Austern, A.J.C. Bik, J.C. Dehnert, I. Horn, N. Leiser, G. Czajkowski, Pregel: a system for large-scale graph processing, in *SIGMOD*, 2010
44. M. Zaharia, M. Chowdhury, M.J. Franklin, S. Shenker, I. Stoica, Spark: cluster computing with working sets, in *HotCloud*, 2010
45. M. Odersky, L. Spoon, B. Venners, *Programming in Scala: A Comprehensive Step-by-Step Guide* (Artima Inc., Walnut Creek, 2011)

46. B. Hindman, A. Konwinski, M. Zaharia, A. Ghodsi, A.D. Joseph, R.H. Katz, S. Shenker, I. Stoica, Mesos: a platform for fine-grained resource sharing in the data center, in *NSDI*, 2011

47. M. Zaharia, D. Borthakur, J.S. Sarma, K. Elmeleegy, S. Shenker, I. Stoica, Delay scheduling: a simple technique for achieving locality and fairness in cluster scheduling, in *EuroSys* (2010), pp. 265–278

48. K. Shvachko, H. Kuang, S. Radia, R. Chansler, The Hadoop distributed file system, in *MSST*, 2010

49. M. Armbrust, R.S. Xin, C. Lian, Y. Huai, D. Liu, J.K. Bradley, X. Meng, T. Kaftan, M.J. Franklin, A. Ghodsi, M. Zaharia, Spark SQL: relational data processing in Spark, in *SIGMOD*, 2015

50. E.R. Sparks, A. Talwalkar, V. Smith, J. Kottalam, X. Pan, J.E. Gonzalez, M.J. Franklin, M.I. Jordan, T. Kraska, MLI: an API for distributed machine learning, in *ICDM*, 2013

51. J.E. Gonzalez, R.S. Xin, A. Dave, D. Crankshaw, M.J. Franklin, I. Stoica, GraphX: graph processing in a distributed dataflow framework, in *OSDI*, 2014

52. A. Alexandrov, R. Bergmann, S. Ewen, J.-C. Freytag, F. Hueske, A. Heise, O. Kao, M. Leich, U. Leser, V. Markl, F. Naumann, M. Peters, A. Rheinländer, M.J. Sax, S. Schelter, M. Höger, K. Tzoumas, D. Warneke, The stratosphere platform for big data analytics. VLDB J. **23**(6), 939–964 (2014)

53. A. Alexandrov, D. Battré, S. Ewen, M. Heimel, F. Hueske, O. Kao, V. Markl, E. Nijkamp, D. Warneke, Massively parallel data analysis with PACTs on nephele. Proc. VLDB Endowment **3**(2), 1625–1628 (2010)

54. D. Battré et al., Nephele/PACTs: a programming model and execution framework for web-scale analytical processing, in *SoCC*, 2010

55. P.G. Selinger, M.M. Astrahan, D.D. Chamberlin, R.A. Lorie, T.G. Price, Access path selection in a relational database management system, in *SIGMOD*, 1979

56. A. Heise, A. Rheinländer, M. Leich, U. Leser, F. Naumann, Meteor/Sopremo: an extensible query language and operator model, in *VLDB Workshops*, 2012

57. V.R. Borkar, M.J. Carey, R. Grover, N. Onose, R. Vernica, Hyracks: a flexible and extensible foundation for data-intensive computing, in *ICDE*, 2011

58. A. Behm, V.R. Borkar, M.J. Carey, R. Grover, C. Li, N. Onose, R. Vernica, A. Deutsch, Y. Papakonstantinou, V.J. Tsotras, ASTERIX: towards a scalable, semistructured data platform for evolving-world models. Distrib. Parallel Databases **29**(3), 185–216 (2011)

59. V. Borkar, S. Alsubaiee, Y. Altowim, H. Altwaijry, A. Behm, Y. Bu, M. Carey, R. Grover, Z. Heilbron, Y.-S. Kim, C. Li, P. Pirzadeh, N. Onose, R. Vernica, J. Wen, ASTERIX: an open source system for "Big Data" management and analysis. Proc. VLDB Endowment **5**(2), 1898–1901 (2012)

60. S. Alsubaiee, Y. Altowim, H. Altwaijry, A. Behm, V.R. Borkar, Y. Bu, M.J. Carey, I. Cetindil, M. Cheelangi, K. Faraaz, E. Gabrielova, R. Grover, Z. Heilbron, Y.-S. Kim, C. Li, G. Li, J.M. Ok, N. Onose, P. Pirzadeh, V.J. Tsotras, R. Vernica, J. Wen, T. Westmann, AsterixDB: a scalable, open source BDMS. Proc. VLDB Endowment **7**(14), 1905–1916 (2014)

61. Y. Bu, V.R. Borkar, J. Jia, M.J. Carey, T. Condie, Pregelix: big(ger) graph analytics on a dataflow engine. Proc. VLDB Endowment **8**(2), 161–172 (2014)

62. A. Pavlo, E. Paulson, A. Rasin, D.J. Abadi, D.J. DeWitt, S. Madden, M. Stonebraker, A comparison of approaches to large-scale data analysis, in *SIGMOD*, 2009

63. A. Thusoo, Z. Shao, S. Anthony, D. Borthakur, N. Jain, J.S. Sarma, R. Murthy, H. Liu, Data warehousing and analytics infrastructure at Facebook, in *SIGMOD*, 2010

64. A. Thusoo, Z. Shao, S. Anthony, D. Borthakur, N. Jain, J.S. Sarma, R. Murthy, H. Liu, Data warehousing and analytics infrastructure at Facebook, in *SIGMOD Conference* (2010), pp. 1013–1020

65. B. Saha, H. Shah, S. Seth, G. Vijayaraghavan, A.C. Murthy, C. Curino, Apache Tez: a unifying framework for modeling and building data processing applications, in *SIGMOD*, 2015

66. V.K. Vavilapalli, A.C. Murthy, C. Douglas, S. Agarwal, M. Konar, R. Evans, T. Graves, J. Lowe, H. Shah, S. Seth, B. Saha, C. Curino, O. O'Malley, S. Radia, B. Reed, E. Baldeschwieler, Apache Hadoop YARN: yet another resource negotiator, in *SOCC*, 2013

67. M. Kornacker, A. Behm, V. Bittorf, T. Bobrovytsky, C. Ching, A. Choi, J. Erickson, M. Grund, D. Hecht, M. Jacobs, I. Joshi, L. Kuff, D. Kumar, A. Leblang, N. Li, I. Pandis, H. Robinson, D. Rorke, S. Rus, J. Russell, D. Tsirogiannis, S. Wanderman-Milne, M. Yoder, Impala: a modern, open-source SQL engine for Hadoop, in *CIDR*, 2015

68. S. Wanderman-Milne, N. Li, Runtime code generation in Cloudera Impala. IEEE Data Eng. Bull. **37**(1), 31–37 (2014)

69. A. Abouzeid, K. Bajda-Pawlikowski, D.J. Abadi, A. Rasin, A. Silberschatz, HadoopDB: an architectural hybrid of MapReduce and DBMS technologies for analytical workloads. Proc. VLDB Endowment **2**(1), 922–933 (2009)

70. M. Stonebraker, D. Abadi, D. DeWitt, S. Madden, E. Paulson, A. Pavlo, A. Rasin, MapReduce and parallel DBMSs: friends or foes? Commun. ACM **53**(1), 64–71 (2010)

71. H. Choi, J. Son, H. Yang, H. Ryu, B. Lim, S. Kim, Y.D. Chung, Tajo: a distributed data warehouse system on large clusters, in *ICDE*, 2013

72. S. Melnik, A. Gubarev, J.J. Long, G. Romer, S. Shivakumar, M. Tolton, T. Vassilakis, Dremel: interactive analysis of web-scale datasets, Proc. VLDB Endowment **3**(1), 330–339 (2010)

73. D.J. DeWitt, A. Halverson, R.V. Nehme, S. Shankar, J. Aguilar-Saborit, A. Avanes, M. Flasza, J. Gramling, Split query processing in Polybase, in *SIGMOD*, 2013

74. V.R. Gankidi, N. Teletia, J.M. Patel, A. Halverson, D.J. DeWitt, Indexing HDFS data in PDW: splitting the data from the index. Proc. VLDB Endowment **7**(13), 1520–1528 (2014)

75. S. Sakr, E. Pardede (eds.), *Graph Data Management: Techniques and Applications* (IGI Global, Hershey, 2011)

76. S. Sakr, Processing large-scale graph data: a guide to current technology, in *IBM DeveloperWorks* (2013), p. 15

77. A. Khan, S. Elnikety, Systems for big-graphs. Proc. VLDB Endowment **7**(13), 1709–1710 (2014)

78. R. Chen, X. Weng, B. He, M. Yang, Large graph processing in the cloud, in *SIGMOD*, 2010

79. U. Kang, C.E. Tsourakakis, C. Faloutsos, PEGASUS: a peta-scale graph mining system, in *ICDM*, 2009

80. U. Kang, H. Tong, J. Sun, C.-Y. Lin, C. Faloutsos, GBASE: a scalable and general graph management system, in *KDD*, 2011

81. U. Kang, C.E. Tsourakakis, C. Faloutsos, PEGASUS: mining peta-scale graphs. Knowl. Inf. Syst. **27**(2), 303–325 (2011)

82. U. Kang, B. Meeder, C. Faloutsos, Spectral analysis for billion-scale graphs: discoveries and implementation, in *PAKDD*, 2011

83. Z. Khayyat, K. Awara, A. Alonazi, H. Jamjoom, D. Williams, P. Kalnis, Mizan: a system for dynamic load balancing in large-scale graph processing, in *EuroSys*, 2013

84. S. Salihoglu, J. Widom, GPS: a graph processing system, in *SSDBM*, 2013

85. J.E. Gonzalez, Y. Low, H. Gu, D. Bickson, C. Guestrin, PowerGraph: distributed graph-parallel computation on natural graphs, in *OSDI*, 2012

86. A. Kyrola, G.E. Blelloch, C. Guestrin, GraphChi: large-scale graph computation on just a PC, in *OSDI*, 2012

87. Y. Low, J. Gonzalez, A. Kyrola, D. Bickson, C. Guestrin, J.M. Hellerstein, Distributed GraphLab: a framework for machine learning in the cloud. Proc. VLDB Endowment **5**(8), 716–727 (2012)

88. B. Shao, H. Wang, Y. Li, Trinity: a distributed graph engine on a memory cloud, in *SIGMOD*, 2013

89. G. Wang, W. Xie, A. Demers, J. Gehrke, Asynchronous large-scale graph processing made easy, in *CIDR*, 2013

90. P. Stutz, A. Bernstein, W.W. Cohen, Signal/collect: graph algorithms for the (semantic) web, in *International Semantic Web Conference (1)*, 2010

91. L.G. Valiant, A bridging model for parallel computation. Commun. ACM **33**(8), 103–111 (1990)

92. W.D. Clinger, Foundations of actor semantics. Technical report, Cambridge (1981)

93. Y. Tian, A. Balmin, S.A. Corsten, S. Tatikonda, J. McPherson, From "think like a vertex" to "think like a graph". Proc. VLDB Endowment **7**(3), 193–204 (2013)

94. A. Dave, A. Jindal, L.E. Li, R. Xin, J. Gonzalez, M. Zaharia, GraphFrames: an integrated API for mixing graph and relational queries, in *Proceedings of the Fourth International Workshop on Graph Data Management Experiences and Systems* (ACM, New York, 2016), p. 2

95. M. Junghanns, A. Petermann, K. Gómez, E. Rahm, Gradoop: scalable graph data management and analytics with Hadoop (2015). Preprint. arXiv:1506.00548

96. M. Kricke, E. Peukert, E. Rahm, Graph data transformations in Gradoop, in *BTW 2019*, 2019

97. N. Francis, A. Green, P. Guagliardo, L. Libkin, T. Lindaaker, V. Marsault, S. Plantikow, M. Rydberg, P. Selmer, A, Taylor, Cypher: an evolving query language for property graphs, in *Proceedings of the 2018 International Conference on Management of Data* (ACM, New York, 2018), pp. 1433–1445

98. M. Junghanns, M. Kießling, A. Averbuch, A. Petermann, E. Rahm, Cypher-based graph pattern matching in Gradoop, in *Proceedings of the Fifth International Workshop on Graph Data-management Experiences & Systems* (ACM, New York, 2017), p. 3

99. M. Junghanns, M. Kießling, N. Teichmann, K. Gómez, A. Petermann, E. Rahm, Declarative and distributed graph analytics with Gradoop. Proc. VLDB Endowment **11**(12), 2006–2009 (2018)

100. W.-S. Han, S. Lee, K. Park, J.-H. Lee, M.-S. Kim, J. Kim, H. Yu, TurboGraph: a fast parallel graph engine handling billion-scale graphs in a single PC, in *KDD*, 2013

101. D. Yan, J. Cheng, Y. Lu, W. Ng, Blogel: a block-centric framework for distributed computation on real-world graphs. Proc. VLDB Endowment **7**(14), 1981–1992 (2014)

102. World Wide Web Consortium. RDF 1.1 Primer (2014)

103. F. Manola, E. Miller. RDF Primer, February 2004. http://www.w3.org/TR/2004/REC-rdf-primer-20040210/

104. E. Prud'hommeaux, A. Seaborne, SPARQL Query Language for RDF, W3C Recommendation, January 2008. http://www.w3.org/TR/rdf-sparql-query/

105. Z. Kaoudi, I. Manolescu, RDF in the clouds: a survey. VLDB J. **24**(1), 67–91 (2015)

106. M. Wylot, M. Hauswirth, P. Cudré-Mauroux, S. Sakr, RDF data storage and query processing schemes: a survey. ACM Comput. Surv. **51**(4), 84:1–84:36 (2018)

107. V. Khadilkar, M. Kantarcioglu, B.M. Thuraisingham, P. Castagna, Jena-HBase: a distributed, scalable and efficient RDF triple store, in *Proceedings of the ISWC 2012 Posters & Demonstrations Track*, Boston, 11–15 November 2012

108. R. Punnoose, A. Crainiceanu, D. Rapp, SPARQL in the cloud using Rya. Inf. Syst. **48**, 181–195 (2015)

109. A. Aranda-Andújar, F. Bugiotti, J. Camacho-Rodríguez, D. Colazzo, F. Goasdoué, Z. Kaoudi, I. Manolescu, AMADA: web data repositories in the amazon cloud, in *21st ACM International Conference on Information and Knowledge Management, CIKM'12*, Maui, 29 October–02 November 2012, pp. 2749–2751

110. G. Ladwig, A. Harth, Cumulusrdf: linked data management on nested key-value stores, in *The 7th International Workshop on Scalable Semantic Web Knowledge Base Systems (SSWS 2011)*, vol. 30 (2011)

111. A. Lakshman, P. Malik, Cassandra: a decentralized structured storage system. SIGOPS Oper. Syst. Rev. **44**(2), 35–40 (2010)

112. R. Mutharaju, S. Sakr, A. Sala, P. Hitzler, D-SPARQ: distributed, scalable and efficient RDF query engine, in *Proceedings of the ISWC 2013 Posters & Demonstrations Track*, Sydney, 23 October 2013, pp. 261–264

113. J. Huang, D.J. Abadi, K. Ren, Scalable SPARQL querying of large RDF graphs. Proc. VLDB Endowment **4**(11), 1123–1134 (2011)

114. N. Papailiou, I. Konstantinou, D. Tsoumakos, P. Karras, N. Koziris, H2RDF+: high-performance distributed joins over large-scale RDF graphs, in *2013 IEEE International Conference on Big Data* (IEEE, Piscataway, 2013), pp. 255–263

115. J. Huang, D.J. Abadi, K. Ren, Scalable SPARQL querying of large RDF graphs. Proc. VLDB Endowment **4**(11), 1123–1134 (2011)

116. A. Abouzied, K. Bajda-Pawlikowski, J. Huang, D.J. Abadi, A. Silberschatz, HadoopDB in action: building real world applications, in *Proceedings of the ACM SIGMOD International Conference on Management of Data, SIGMOD 2010*, Indianapolis, 6–10 June 2010, pp. 1111–1114

117. T. Neumann, G. Weikum, RDF-3X: a RISC-style engine for RDF. Proc. VLDB Endowment **1**(1), 647–659 (2008)

118. F. Goasdoué, Z. Kaoudi, I. Manolescu, J.-A. Quiané-Ruiz, S. Zampetakis, CliqueSquare: flat plans for massively parallel RDF queries, in *31st IEEE International Conference on Data Engineering, ICDE 2015*, Seoul, 13–17 April 2015, pp. 771–782

119. B. Djahandideh, F. Goasdoué, Z. Kaoudi, I. Manolescu, J.-A. Quiané-Ruiz, S. Zampetakis, CliqueSquare in action: flat plans for massively parallel RDF queries, in *31st IEEE International Conference on Data Engineering, ICDE 2015*, Seoul, 13–17 April 2015, pp. 1432–1435

120. A. Schätzle, M. Przyjaciel-Zablocki, T. Hornung, G. Lausen, PigSPARQL: a SPARQL query processing baseline for big data, in *Proceedings of the ISWC 2013 Posters & Demonstrations Track*, Sydney, 23 October 2013, pp. 241–244

121. C. Olston, B. Reed, U. Srivastava, R. Kumar, A. Tomkins, Pig Latin: a not-so-foreign language for data processing, in *Proceedings of the ACM SIGMOD International Conference on Management of Data, SIGMOD 2008*, Vancouver, 10–12 June 2008, pp. 1099–1110

122. P. Ravindra, H. Kim, K. Anyanwu, An intermediate algebra for optimizing RDF graph pattern matching on MapReduce, in *The Semantic Web: Research and Applications - 8th Extended Semantic Web Conference, ESWC 2011. Proceedings, Part II*, Heraklion, Crete, 29 May–2 June 2011, pp. 46–61

123. H. Kim, P. Ravindra, K. Anyanwu, Optimizing RDF(S) queries on cloud platforms, in *22nd International World Wide Web Conference, WWW '13*, Rio de Janeiro, 13–17 May 2013, Companion Volume (2013), pp. 261–264

124. A. Schätzle, M. Przyjaciel-Zablocki, S. Skilevic, G. Lausen, S2RDF: RDF querying with SPARQL on Spark. CoRR (2015), abs/1512.07021

125. D.J. Abadi, A. Marcus, S.R. Madden, K. Hollenbach, Scalable semantic web data management using vertical partitioning, in *Proceedings of the 33rd International Conference on Very Large Data Bases*, VLDB Endowment (2007), pp. 411–422

126. P. Valduriez, Join indices. ACM Trans. Database Syst. **12**(2), 218–246 (1987)

127. P.A. Bernstein, D.-M.W. Chiu, Using semi-joins to solve relational queries. J. ACM **28**(1), 25–40 (1981)

128. X. Chen, H. Chen, N. Zhang, S. Zhang, SparkRDF: elastic discreted RDF graph processing engine with distributed memory, in *Proceedings of the ISWC 2014 Posters & Demonstrations Track a track within the 13th International Semantic Web Conference, ISWC 2014*, Riva del Garda, 21 October 2014, pp. 261–264

129. X. Chen, H. Chen, N. Zhang, S. Zhang, SparkRDF: elastic discreted RDF graph processing engine with distributed memory, in *IEEE/WIC/ACM International Conference on Web Intelligence and Intelligent Agent Technology, WI-IAT 2015*, Volume I, Singapore, 6–9 December 2015, pp. 292–300

130. A. Schätzle, M. Przyjaciel-Zablocki, T. Berberich, G. Lausen, S2X: graph-parallel querying of RDF with GraphX, in *1st International Workshop on Big-Graphs Online Querying (Big-O(Q)*, 2015

131. E.L. Goodman, D. Grunwald, Using vertex-centric programming platforms to implement SPARQL queries on large graphs, in *Proceedings of the 4th Workshop on Irregular Applications: Architectures and Algorithms*, IA3 '14 (IEEE Press, Piscataway, 2014), pp. 25–32

132. H. Naacke, O. Curé, B. Amann, SPARQL query processing with Apache Spark (2016). CoRR, abs/1604.08903

133. K. Zeng, J. Yang, H. Wang, B. Shao, Z. Wang, A distributed graph engine for web scale RDF data, in *Proceedings of the 39th International Conference on Very Large Data Bases*. VLDB Endowment (2013), pp. 265–276

134. P. Stutz, M. Verman, L. Fischer, A. Bernstein, TripleRush: a fast and scalable triple store, in *SSWS@ ISWC* (2013), pp. 50–65

135. P. Stutz, B. Paudel, M. Verman, A. Bernstein, Random walk TripleRush: asynchronous graph querying and sampling, in *Proceedings of the 24th International Conference on World Wide Web, WWW 2015*, Florence, 18–22 May 2015, pp. 1034–1044

136. P. Stutz, A. Bernstein, W. Cohen, Signal/collect: graph algorithms for the (semantic) web, in *International Semantic Web Conference* (Springer, Berlin, 2010), pp. 764–780

137. R. Harbi, I. Abdelaziz, P. Kalnis, N. Mamoulis, Evaluating SPARQL queries on massive RDF datasets. Proc. VLDB Endowment **8**(12), 1848–1851 (2015)

138. R. Al-Harbi, I. Abdelaziz, P. Kalnis, N. Mamoulis, Y. Ebrahim, M. Sahli, Accelerating SPARQL queries by exploiting hash-based locality and adaptive partitioning. VLDB J. **25**(3), 355–380 (2016)

139. S. Gurajada, S. Seufert, I. Miliaraki, M. Theobald, TriAD: a distributed shared-nothing RDF engine based on asynchronous message passing, in *International Conference on Management of Data, SIGMOD 2014*, Snowbird, 22–27 June 2014, pp. 289–300

140. L. Galárraga, K. Hose, R. Schenkel, Partout: a distributed engine for efficient RDF processing, in *23rd International World Wide Web Conference, WWW '14*, Seoul, 7–11 April 2014, Companion Volume, pp. 267–268

141. T. Neumann, G. Weikum, The RDF-3X engine for scalable management of RDF data. VLDB J. **19**(1), 91–113 (2010)

142. M. Hammoud, D.A. Rabbou, R. Nouri, S.-M.-R. Beheshti, S. Sakr, DREAM: distributed RDF engine with adaptive query planner and minimal communication. Proc. VLDB Endowment **8**(6), 654–665 (2015)

143. A. Hasan, M. Hammoud, R. Nouri, S. Sakr, DREAM in action: a distributed and adaptive RDF system on the cloud, in *Proceedings of the 25th International Conference on World Wide Web, WWW 2016*, Montreal, 11–15 April 2016, Companion Volume, pp. 191–194

144. L. Cheng, S. Kotoulas, Scale-out processing of large RDF datasets. IEEE Trans. Big Data **1**(4), 138–150 (2015)

145. M. Wylot, P. Cudré-Mauroux, DiploCloud: efficient and scalable management of RDF data in the cloud. IEEE Trans. Knowl. Data Eng. **28**(3), 659–674 (2016)

146. P. Zikopoulos, C. Eaton et al., *Understanding Big Data: Analytics for Enterprise Class Hadoop and Streaming Data* (McGraw-Hill Osborne Media, New York, 2011)

147. K. Ashton et al., That 'Internet of things' thing. RFID J. **22**(7), 97–114 (2009)

148. N. Marz, J. Warren, *Big Data: Principles and Best Practices of Scalable Realtime Data Systems* (Manning Publications Co., Shelter Island, 2015)

149. T. Condie, N. Conway, P. Alvaro, J.M. Hellerstein, K. Elmeleegy, R. Sears, MapReduce online, in *NSDI*, 2010

150. T. Condie, N. Conway, P. Alvaro, J.M. Hellerstein, J. Gerth, J. Talbot, K. Elmeleegy, R. Sears, Online aggregation and continuous query support in MapReduce, in *SIGMOD*, 2010

151. D. Logothetis, K. Yocum, Ad-hoc data processing in the cloud. Proc. VLDB Endowment **1**(2), 1472–1475 (2008)

152. P. Bhatotia, A. Wieder, R. Rodrigues, U.A. Acar, R. Pasquini, Incoop: MapReduce for incremental computations, in *SOCC*, 2011

153. A.M. Aly, A. Sallam, B.M. Gnanasekaran, L.-V. Nguyen-Dinh, W.G. Aref, M. Ouzzaniy, A. Ghafoor, M3: stream processing on main-memory MapReduce, in *ICDE*, 2012

154. V. Kumar, H. Andrade, B. Gedik, K.-L. Wu, DEDUCE: at the intersection of MapReduce and stream processing, in *EDBT* (2010), pp. 657–662

155. S. Sakr, An introduction to InfoSphere Streams: a platform for analyzing big data in motion. IBM DeveloperWorks, 2013. http://www.ibm.com/developerworks/library/bd-streamsintro/index.html

156. S. Loesing, M. Hentschel, T. Kraska, D. Kossmann, Stormy: an elastic and highly available streaming service in the cloud, in *EDBT/ICDT Workshops*, 2012

157. H. Balakrishnan, M. Frans Kaashoek, D.R. Karger, R. Morris, I. Stoica, Looking up data in p2p systems. Commun. ACM **46**(2), 43–48 (2003)

158. L. Neumeyer, B. Robbins, A. Nair, A. Kesari, S4: distributed stream computing platform, in *ICDMW*, 2010

159. B. Gedik, H. Andrade, K.-L. Wu, P.S. Yu, M. Doo, SPADE: the system S declarative stream processing engine, in *SIGMOD*, 2008

160. M. Armbrust, T. Das, J. Torres, B. Yavuz, S. Zhu, R. Xin, A. Ghodsi, I. Stoica, M. Zaharia, Structured streaming: a declarative API for real-time applications in Apache Spark, in *SIGMOD*, 2018

161. J. Kreps, N. Narkhede, J. Rao et al., Kafka: a distributed messaging system for log processing, in *Proceedings of the NetDB*, 2011

162. S. Kulkarni, N. Bhagat, M. Fu, V. Kedigehalli, C. Kellogg, S. Mittal, J.M. Patel, K. Ramasamy, S. Taneja, Twitter Heron: stream processing at scale, in *SIGMOD*, 2015

163. G. De Francisci Morales, A. Bifet, Samoa: scalable advanced massive online analysis. J. Mach. Learn. Res. **16**(1), 149–153 (2015)

164. A. Gates, O. Natkovich, S. Chopra, P. Kamath, S. Narayanam, C. Olston, B. Reed, S. Srinivasan, U. Srivastava, Building a highlevel dataflow system on top of MapReduce: the Pig experience. Proc. VLDB Endowment **2**(2), 1414–1425 (2009)

165. A. Gates, *Programming Pig* (O'Reilly Media, Sebastopol, 2011)

166. C. Chambers, A. Raniwala, F. Perry, S. Adams, R.R. Henry, R. Bradshaw, N. Weizenbaum, FlumeJava: easy, efficient data-parallel pipelines, in *PLDI*, 2010

167. D. Wu, L. Zhu, X. Xu, S. Sakr, D. Sun, Q. Lu, A pipeline framework for heterogeneous execution environment of big data processing. IEEE Softw. **33**, 60–67 (2016)

168. R. Elshawi, S. Sakr, D. Talia, P. Trunfio, Big data systems meet machine learning challenges: towards big data science as a service. Big Data Res. **14**, 1–11 (2018)

169. D. Michie, D.J. Spiegelhalter, C.C. Taylor et al., *Machine Learning. Neural and Statistical Classification*, vol. 13 (Ellis Horwood, London, 1994)

170. S. Owen, *Mahout in Action* (Manning Publications Co., Shelter Island, 2012)

171. X. Meng, J. Bradley, B. Yavuz, E. Sparks, S. Venkataraman, D. Liu, J. Freeman, D.B. Tsai, M. Amde, S. Owen et al., MLlib: machine learning in Apache Spark. J. Mach. Learn. Res. **17**(1), 1235–1241 (2016)

172. M. Stonebraker, P. Brown, A. Poliakov, S. Raman, The architecture of SciDB, in *International Conference on Scientific and Statistical Database Management* (Springer, Berlin, 2011), pp. 1–16

173. X. Li, B. Cui, Y. Chen, W. Wu, C. Zhang, MLog: towards declarative in-database machine learning. Proc. VLDB Endowment **10**(12), 1933–1936 (2017)

174. P.G. Brown, Overview of SciDB: large scale array storage, processing and analysis, in *Proceedings of the 2010 ACM SIGMOD International Conference on Management of Data* (ACM, New York, 2010), pp. 963–968

175. J.M. Hellerstein, C. Ré, F. Schoppmann, D.Z. Wang, E. Fratkin, A. Gorajek, K.S. Ng, C. Welton, X. Feng, K. Li et al., The MADlib analytics library: or MAD skills, the SQL. Proc. VLDB Endowment **5**(12), 1700–1711 (2012)

176. S. Das, Y. Sismanis, K.S. Beyer, R. Gemulla, P.J. Haas, J. McPherson, Ricardo: integrating R and Hadoop, in *Proceedings of the 2010 ACM SIGMOD International Conference on Management of Data* (ACM, New York, 2010), pp. 987–998

177. S. Venkataraman, Z. Yang, D. Liu, E. Liang, H. Falaki, X. Meng, R. Xin, A. Ghodsi, M. Franklin, I. Stoica et al., SparkR: scaling R programs with Spark, in *Proceedings of the 2016 International Conference on Management of Data* (ACM, New York, 2016), pp. 1099–1104

178. S. Leo, G. Zanetti, Pydoop: a Python MapReduce and HDFS API for Hadoop, in *Proceedings of the 19th ACM International Symposium on High Performance Distributed Computing* (ACM, New York, 2010), pp. 819–825

179. AzureML Team. AzureML: anatomy of a machine learning service, in *Conference on Predictive APIs and Apps* (2016), pp. 1–13

180. B. Huang, S. Babu, J. Yang, Cumulon: optimizing statistical data analysis in the cloud, in *Proceedings of the 2013 ACM SIGMOD International Conference on Management of Data* (ACM, New York, 2013), pp. 1–12

181. M. Boehm, M.W. Dusenberry, D. Eriksson, A.V. Evfimievski, F.M. Manshadi, N. Pansare, B. Reinwald, F.R. Reiss, P. Sen, A.C. Surve et al., SystemML: declarative machine learning on Spark. Proc. VLDB Endowment **9**(13), 1425–1436 (2016)
182. S. Schelter, A. Palumbo, S. Quinn, S. Marthi, A. Musselman, Samsara: declarative machine learning on distributed dataflow systems, in *NIPS Workshop ML Systems*, 2016
183. T. Kraska, A. Talwalkar, J.C. Duchi, R. Griffith, M.J. Franklin, M.I. Jordan, MLbase: a distributed machine-learning system, in *CIDR*, 2013
184. M. Weimer, T. Condie, R. Ramakrishnan et al., Machine learning in ScalOps, a higher order cloud computing language, in *NIPS 2011 Workshop on Parallel and Large-Scale Machine Learning (BigLearn)*, vol. 9 (2011), pp. 389–396
185. V. Borkar, M. Carey, R. Grover, N. Onose, R. Vernica, Hyracks: a flexible and extensible foundation for data-intensive computing, in *2011 IEEE 27th International Conference on Data Engineering* (IEEE, Piscataway, 2011), pp. 1151–1162
186. E.R. Sparks, S. Venkataraman, T. Kaftan, M.J. Franklin, B. Recht, Keystoneml: optimizing pipelines for large-scale advanced analytics, in *2017 IEEE 33rd International Conference on Data Engineering (ICDE)* (IEEE, Piscataway, 2017), pp. 535–546
187. Y. LeCun, Y. Bengio, G. Hinton, Deep learning. Nature **521**(7553), 436–444 (2015)
188. I. Goodfellow, Y. Bengio, A. Courville, *Deep Learning* (MIT Press, Cambridge, 2016)
189. A. Krizhevsky, I. Sutskever, G.E. Hinton, ImageNet classification with deep convolutional neural networks, in *Advances in Neural Information Processing Systems* (2012), pp. 1097–1105
190. R. Collobert et al., Natural language processing (almost) from scratch. J. Mach. Learn. Res. **12**, 2493–2537 (2011)
191. G. Hinton et al., Deep neural networks for acoustic modeling in speech recognition: the shared views of four research groups. IEEE Signal Process. Mag. **29**(6), 82–97 (2012)
192. Y. Bengio et al., Learning deep architectures for AI. Found. Trends Mach. Learn. **2**(1), 1–127 (2009)
193. M. Abadi et al., TensorFlow: a system for large-scale machine learning, in *OSDI*, vol. 16 (2016), pp. 265–283
194. D. Baylor, E. Breck, H.-T. Cheng, N. Fiedel, C.Y. Foo, Z. Haque, S. Haykal, M. Ispir, V. Jain, L. Koc et al., TFX: a TensorFlow-based production-scale machine learning platform, in *Proceedings of the 23rd ACM SIGKDD International Conference on Knowledge Discovery and Data Mining* (ACM, New York, 2017), pp. 1387–1395
195. J. Bergstra et al., Theano: a CPU and GPU math compiler in Python, in *Proceedings of 9th Python in Science Conference*, vol. 1, 2010
196. T. Chen et al., MXNet: a flexible and efficient machine learning library for heterogeneous distributed systems (2015). Preprint. arXiv:1512.01274
197. A. Paszke, S. Gross, S. Chintala, G. Chanan, E. Yang, Z. DeVito, Z. Lin, A. Desmaison, L. Antiga, A. Lerer, Automatic differentiation in PyTorch (2017)
198. S. Tokui, K. Oono, S. Hido, J. Clayton, Chainer: a next-generation open source framework for deep learning, in *NIPS Workshops*, 2015
199. S. Lohr, The age of big data. New York Times, 11, 2012
200. V. Mayer-Schönberger, K. Cukier, *Big Data: A Revolution that Will Transform How We Live, Work, and Think* (Houghton Mifflin Harcourt, Boston, 2013)
201. H.E. Schaffer, X as a service, cloud computing, and the need for good judgment. IT Prof. **11**(5), 4–5 (2009)
202. D. Delen, H. Demirkan, Data, information and analytics as services. Decis. Support Syst. **55**(1), 359–363 (2013)
203. M. Baker, Data science: industry allure. Nature **520**, 253–255 (2015)
204. F. Provost, T. Fawcett, Data science and its relationship to big data and data-driven decision making. Big Data **1**(1), 51–59 (2013)
205. A. Labrinidis, H.V. Jagadish, Challenges and opportunities with big data. Proc. VLDB Endowment **5**(12), 2032–2033 (2012)

206. H.V. Jagadish, J. Gehrke, A. Labrinidis, Y. Papakonstantinou, J.M. Patel, R. Ramakrishnan, C. Shahabi, Big data and its technical challenges. Commun. ACM **57**(7), 86–94 (2014)
207. D. Abadi, S. Babu, F. Ozcan, I. Pandis, SQL-on-Hadoop systems. Proc. VLDB Endowment **8**(12), 2050–2061 (2015)
208. S. Sakr, S. Elnikety, Y. He, G-SPARQL: a hybrid engine for querying large attributed graphs, in *CIKM* (2012), pp. 335–344
209. Y. Guo, A.L. Varbanescu, A. Iosup, C. Martella, T.L. Willke, Benchmarking graph-processing platforms: a vision, in *ICPE*, 2014
210. A. Barnawi, O. Batarfi, S.-M.-R. Beheshti, R. El Shawi, A.G. Fayoumi, R. Nouri, S. Sakr, On characterizing the performance of distributed graph computation platforms, in *TPCTC*, 2014
211. O. Batarfi, R. El Shawi, A.G. Fayoumi, R. Nouri, S.-M.-R. Beheshti, A. Barnawi, S. Sakr, Large scale graph processing systems: survey and an experimental evaluation. Clust. Comput. **18**(3), 1189–1213 (2015)
212. M. Han, K. Daudjee, K. Ammar, M. Tamer Özsu, X. Wang, T. Jin, An experimental comparison of Pregel-like graph processing systems. Proc. VLDB Endowment **7**(12), 1047–1058 (2014)
213. Y. Lu, J. Cheng, D. Yan, H. Wu, Large-scale distributed graph computing systems: an experimental evaluation. Proc. VLDB Endowment **8**(3), 281–292 (2014)
214. Y. Guo, M. Biczak, A.L. Varbanescu, A. Iosup, C. Martella, T.L. Willke, How well do graph-processing platforms perform? An empirical performance evaluation and analysis, in *IPDPS*, 2014
215. M. Li, J. Tan, Y. Wang, L. Zhang, V. Salapura, SparkBench: a comprehensive benchmarking suite for in memory data analytic platform Spark, in *Proceedings of the 12th ACM International Conference on Computing Frontiers, CF'15*, Ischia, 18–21 May 2015, pp. 53:1–53:8
216. M. Capota, T. Hegeman, A. Iosup, A. Prat-Pérez, O. Erling, P.A. Boncz, Graphalytics: a big data benchmark for graph-processing platforms, in *Proceedings of the Third International Workshop on Graph Data Management Experiences and Systems, GRADES 2015*, Melbourne, 31 May–4 June 2015, pp. 7:1–7:6
217. O. Batarfi, R. El Shawi, A.G. Fayoumi, R. Nouri, A. Barnawi, S. Sakr et al., Large scale graph processing systems: survey and an experimental evaluation. Clust. Comput. **18**(3), 1189–1213 (2015)
218. V. Aluko, S. Sakr, Big SQL systems: an experimental evaluation. Clust. Comput. **22**(4), 1347–1377 (2019)
219. N. Mahmoud, Y. Essam, R. El Shawi, S. Sakr, DLBench: an experimental evaluation of deep learning frameworks, in *2019 IEEE International Congress on Big Data, BigData Congress 2019*, Milan, 8–13 July 2019, pp. 149–156
220. E. Shahverdi, A. Awad, S. Sakr, Big stream processing systems: an experimental evaluation, in *2019 IEEE 35th International Conference on Data Engineering Workshops (ICDEW)* (IEEE, Piscataway, 2019), pp. 53–60
221. I. Gog, M. Schwarzkopf, N. Crooks, M.P. Grosvenor, A. Clement, S. Hand, Musketeer: all for one, one for all in data processing systems, in *EuroSys* (2015), pp. 2:1–2:16
222. D. Agrawal, M. Lamine Ba, L. Berti-Equille, S. Chawla, A. Elmagarmid, H. Hammady, Y. Idris, Z. Kaoudi, Z. Khayyat, S. Kruse, M. Ouzzani, P. Papotti, J.-A. Quiané-Ruiz, N. Tang, M.J. Zaki, Rheem: enabling multi-platform task execution, in *SIGMOD Conference*, 2016
223. N. Huijboom, T. Van den Broek, Open data: an international comparison of strategies. Eur. J. ePractice **12**(1), 4–16 (2011)
224. M. Balazinska, B. Howe, D. Suciu, Data markets in the cloud: an opportunity for the database community. Proc. VLDB Endowment **4**(12), 1482–1485 (2011)
225. R. El Shawi, M. Maher, S. Sakr, Automated machine learning: state-of-the-art and open challenges (2019). CoRR, abs/1906.02287

226. H. Miao, A. Li, L.S. Davis, A. Deshpande, ModelHub: deep learning lifecycle management, in *2017 IEEE 33rd International Conference on Data Engineering (ICDE)* (IEEE, Piscataway, 2017), pp. 1393–1394
227. M. Vartak, H. Subramanyam, W.-E. Lee, S. Viswanathan, S. Husnoo, S. Madden, M. Zaharia, Model DB: a system for machine learning model management, in *Proceedings of the Workshop on Human-In-the-Loop Data Analytics* (ACM, New York, 2016), p. 14
228. P. Bailis, K. Olukotun, C. Ré, M. Zaharia, Infrastructure for usable machine learning: the Stanford DAWN Project (2017). Preprint. arXiv:1705.07538

.

Printed in the United States
by Baker & Taylor Publisher Services